TOWARDS 2001

ABOUT THE AUTHORS

MALCOLM ABRAMS is a magazine consultant and book author. He has worked with major publishing companies as a writer, editor, and editorial director. A native of Toronto, he lives in Manhattan.

HARRIET BERNSTEIN is a former reporter for *Money* magazine and a graduate of the Radcliffe Publishing Program. A recovering technophobe Ms. Bernstein happened onto *Towards 2001* by accident. Although now quite taken by the subject, she still doesn't own a TV.

TOWARDS 2001

A CONSUMER'S GUIDE TO THE 21ST CENTURY

Malcolm Abrams
and Harriet Bernstein

A division of HarperCollins *Publishers*

AN ANGUS & ROBERTSON BOOK

Angus & Robertson (UK)
16 Golden Square, London W1R 4BN
United Kingdom
Collins/Angus & Robertson Publishers Australia
Unit 4, Eden Park, 31 Waterloo Road,
North Ryde, NSW, Australia 2113 and
William Collins Publishers Ltd
31 View Road, Glenfield, Auckland 10,
New Zealand

This book is copyright.
Apart from any fair dealing for the
purposes of private study, research,
criticism or review, as permitted
under the Copyright Act, no part may be
reproduced by any process without
written permission. Inquiries should
be addressed to the publishers.

First published in the United States by
Penguin Books USA Inc in 1989
First published in the United Kingdom by
Angus & Robertson (UK) in 1990
First published in Australia by
Collins/Angus & Robertson Australia in 1990

Copyright © Malcolm Abrams and Harriet Bernstein 1989

Many of the trade names in this book are
registered trademarks; they are protected by
United States and international trademark laws

Typeset in Great Britain by New Faces, Bedford
Printed in Finland

British Library Cataloguing in Publication Data
   Abrams, Malcolm
   Towards 2001
   1. Inventions
   I. Title II. Bernstein, Harriet
608
ISBN 0 207 16657 9

The right of Malcolm Abrams and Harriet Bernstein
to be identified as authors of this work has been
asserted by them in accordance with the Copyright,
Designs and Patents Act 1988

For Paul and Aaron, Gabrielle and Max
May the future be as Good and Bright

ACKNOWLEDGMENTS

A book that covers this much material could never be done without the help of a strong body of researchers. We were exceptionally fortunate to receive help from friends, colleagues, and new faces that came to us through a healthy grapevine. Our deepest thanks go to Fritz Bernstein, Grace Bennett, Jane Berryman, Diane Burley, Suzanne Carmick, Peter Cerbone, Wendy Cole, Chris Collins, Sue Crystal, Jennie Danowski, Kris DiLorenzo, Fonda Duvanel, Sarasue French, Annette Foglino, Tom Garafola, Diane Giles, Ira Hellman, Roger Jennings, Easy Klein, Miriam Leuchter, Matthew Levine, Chris Levite, Elizabeth MacDonald, Chris Markle, Ann Whipple Marr, Julie Moline, Lisa Murray, Didi Pershouse, Judy Sandra, George Schaub, Chris Secrest, Jacqueline Smith, Pam Supplee, Divya Symmers, Linn Varney, Valerie Vaz, Monte Williams, Brook Zern. Very special thanks to Norman Meyersohn, our automobile expert, and to Lisa Towle, our most prolific reporter. Thanks also to our agent, Madeleine Morel, our editor, Mindy Werner, her assistant, Janine Steel, as well as to Gina Holloman and Amy Renson, who helped us out at office central.

We are also indebted to the many people in organizations whose purpose is to present the marketplace with great ideas and inventions discovered in their worlds. There was Pam Michaelson of the *International New Product Newsletter*, Al Kaff and Yong Kim of the Cornell News Service, Mark Seitman of NASA, Professor Patrick Purcell of the Massachusetts Institute of Technology Media Lab, Joe Durocher of the University of New Hampshire, Marilyn Miller of Princeton Dental Resources, Judy Sawyer of *Video Magazine*, Peter Moore of *Modern*

Photography, Eric Schrier of *Hippocrates*, Tom Duretsky of *Omni*, Roger Dooley of *Electronic House*, the *New Scientist* magazine, Elaine Stern of Rohla DiClerico, Chuck Wicksome of Food Technology Institute, Deborah Dallinger of the Rochester Institute of Technology, Charles Downs of Michigan State University, Jack Losse of the University of Alabama, Joe Kepneck at Stanford University Tech Transfer, and the many inventors' organizations from Invent America, which runs a contest every year for the best kid's invention, to the National Congress of Inventors' Organizations and all the folks in between: Richard Wantz of the Kessler Corporation, Steve Gnass of the Invention Convention, John Cronin of DirecTech, who publishes *Tech Trans International Technology Transfer Directory*, John Pike of International Business Development, Alan Tratnor of Inventors' Workshop, Intellectual Property Owners, the Small Business Administration, the U.S. Patent Office, Industrial Innovation Service of Canada, Lomar Associates, and others who connected to our grapevine but whose names got lost in the twine. A special appreciation to Blair Newman and Jerome Svigals for making us a little less technophobic and for clarifying all our technology research, and to Susan Sanders of Hammacher Schlemmer, who not only helped us locate terrific inventions around the world but had the graciousness not to laugh at us when we told her what we were going to do.

And where would we be without the cooperation and warm enthusiasm of the many inventors, researchers, public-relations agents, professors, scientists, and corporate executives who spoke to us – many of whom have become new friends – and the consistently amazing support of personal friends and family, particularly Seena Harris-Parker, Nancy and Mildred Abrams, Seymour Bernstein, Bob Swerdling, and Eileen Fisher.

We thank you all from the bottom of our hearts.

CONTENTS

	Introduction	11
	Preface	13
1	**Future Homes and Gardens**	15
2	**Twenty-first Century Communications**	51
3	**Leisure 2001**	77
4	**Health Matters in the Future**	113
5	**Fly Drive into the Future**	155
6	**Revolutionary Shopping**	177
	A Note on Scientific Terms	201
	Index	205

INTRODUCTION

Towards 2001 is a book for consumers. Everything in this book should be in your supermarket, hardware store, pharmacy, department store, or otherwise available by the year 2001. Many of the technologies behind the products in *Towards 2001* are new and developing. So it's doubtful that any one reader is going to know about all of them. For that reason, we have made the book light on scientific and technical talk and heavy on lightness and clarity.

It's our hope that as you read you will get a good idea of what each product does, some idea of how it works, and a clear idea of the need each fulfills. We also hope that you are inspired by the imagination and creativity of the people responsible for the inventions and developments in this book.

As you read along – about flying cars and water walkers, sonic painkillers and painless dental drills, smart cards and smart houses – something else is likely to happen. The products, like pieces in a jigsaw puzzle, start to form pictures. And suddenly, in your mind's eye, you have an exciting vision of what life will be like in the final decade of the twentieth century.

A few words about the products in this book: Not every one will be new to you. From newspapers, magazines, and television, you'll know about some of them. There may even be the occasional item that's being test-marketed in your area or has arrived sooner than expected. Mostly, though, you'll be surprised at all the products coming.

There are serious products, like cholesterol- and fat-free foods, voice-activated computers, radon extractors, and new lenses that will let blind people see. And fun products too: a robot dog, a kiss moisturizer, potato ice cream, and night golf.

In between the fun and the serious are hundreds of useful, entertaining, delicious, invigorating, stimulating, self-improving, time-, energy-, and money-saving products that will make life better.

Think of this book as a kind of window-shopping expedition into the future.

Enjoy picking out what you'll buy tomorrow.

MALCOLM ABRAMS
HARRIET BERNSTEIN

PREFACE

Some of the inventions in this book push forward the boundaries of technology into the twenty-first century, while others required only imagination and a little lateral thinking on the part of their creators. Most of them naturally divide into various categories which are listed in the text:

- convenience
- money-saving
- environment-friendly
- energy-saving
- time-saving
- helping the disabled or those with minor physical impairments such as short sight
- security
- safety
- space-saving
- healthier alternative
- enhancement to an existing invention

And finally, for those inventions that have you bursting out laughing or shaking your head in disbelief

- wacky

It is not usually possible to state with accuracy when these wonderful new concepts might be available in your office or local high street. Problems on production lines during laboratory testing, or the inability to find an interested manufacturer in the first place, may hold up bright ideas for a decade. Conversely, if something catches the public imagination and there is a perceived demand for it, it may be here sooner than anyone expected. Symbols are used in the text to give a rough guide to anticipated availability:

●●● marginally available already, or coming very soon
●●○ available in the mid-1990s
●○○ not available until 2000

1
FUTURE HOMES AND GARDENS

Computerized interior design
Plastic nails
Laminated wood to rival steel
Intercommunicating domestic appliances
Smart House
Microwave clothes dryer
Solar-powered cooker
Cool cooker
Touch-free taps
Intelligent loo
Walking TV
Solar air conditioner
Robot lawn mower
Self-weeding lawn
Noise canceller
Noise meter
Solar lighting system
Solar roofing material
Solar windows
Privacy windows
Window shatter protector
Fire emergency lifeline
Memory metals
Electrical shock hazard protector
VoiceKey
Hand-scanning lock
Robot dog
Non-choking dog collar
'Bark stopper' dog collar
Butler-in-a-Box
Smoke Check badge
Dome homes

COMPUTERIZED INTERIOR DESIGN

  TIME SAVING

No more messing around with paint charts, tape measures and swatches of fabric – a new service will soon be available that should take the guesswork (and the footwork) out of redecorating your home. Called Design and Decorate, it's a computer system that will actually let you see, from any perspective, 3-D pictures of how your room will look with different furniture, layout, colour schemes, wallpaper and upholstery.

A joint effort of American companies Intel and Videodisc Publishing Inc., the system will be capable of storing the equivalent of two hundred thousand pages of text from a catalogue. Using a technology called Digital Video Interactive, various room components can be placed on a screen that could eventually show up to sixteen million colours.

Shops may well charge for this service, but will probably offset the price against whatever you purchase from them for your new-look home.

PLASTIC NAILS

  CONVENIENCE

And while you're changing the look of your home, spare a thought for the nails that bolt your home and furniture together and hold your pictures on the wall. The future, as Dustin Hoffman was so prophetically told in The

Graduate, is one word, 'plastics'. So, watch for plastic nails, coming soon to a DIY store near you . . . .

The New Age nails are surprisingly strong and offer a number of advantages over the metal ones we've been crookedly hammering for the past few centuries. Plastic nails won't rust, they're not magnetic, and they have three times the pull-out resistance of their metal counterparts. In addition, they won't tear your sandpaper or break your blade if you accidentally run your saw over one.

What makes plastic nails so strong is that they melt slightly when they penetrate, thus bonding the nail's surface to the wood. Because they don't rust, they are perfect for any outdoor or underwater job. They take most paints, come in a variety of colours, and can even be custom-tinted. The down-side, unfortunately, is that they have a greater tendency to snap in two. So there may be some jobs for which they aren't suitable.

The new nails have been dreamed up by the clever Japanese, who are already marketing them in the USA to such powerful names as NASA and the Ford Motor Co. At present they aren't much in use by DIY addicts, because to fix them you need a special air gun from the manufacturers, Kotoko Co., and it's rather expensive. But lots of bright new inventions started out high-priced.

LAMINATED WOOD TO RIVAL STEEL

☑ TIME SAVING ☑ MONEY-SAVING
☑ ENVIRONMENT-FRIENDLY

Here's a wood that's tougher than wood – and environmentally-friendly too. Two Canadians, Derek Barnes and Mark Churchland, have come up with a laminated wood that is three times as strong as traditional timber and could do the job of concrete and steel.

The secret lies in the construction of the laminate. Normally laminated wood products, such as plywood, are weaker than 'real' timber. This, Barnes discovered, was because the constituent parts were not assembled with the grain parallel.

So he and his colleague developed Parallam, which is made by meshing together slivers of wood and 'cooking' them to create a hard, solid board. First a log is peeled into sheets, which are then cut into narrow slivers up to 6 feet long. The knot-holes – the weak points of natural wood – are removed, and the slivers covered with resin and cured with microwaves.

Barnes and Churchland see their brainchild competing with reinforced concrete and steel in apartment blocks, warehousing and so on. In modern, open plan-style housing it is both functional and attractive – something that is rarely said of concrete nowadays. Architects like the new material, because it is easier to work with and costs less. A final advantage in our green-conscious society is that Parallam's construction enables it to be used in load-supporting situations that would normally require single logs from large, old trees.

INTERCOMMUNICATING DOMESTIC APPLIANCES

Can you imagine your telephone talking to the programmer on your central heating boiler? Or your TV set flashing up a message to the effect that the milk's boiling over in the kitchen? Such a scenario may not be far off.

Home automation of this kind would be impossible without compatible equipment, so setting up industry standards is an essential prerequisite. Companies in the USA are now close to agreement on just such a standard, called Cebus. Appliances are controlled by signals which come through a loop of wire which is connected to all parts of the system. Similar systems already exist, but are less practical because new wiring has to be installed. Cebus, on the other hand, works via mains power cables, telephone wiring, radio and infra-red signals; in the future, fibre optics will be used.

In Europe, a similar system is being set up. Philips and Matsushita have announced a standard called D2B (Domestic Digital Bus); it won't be restricted to any particular manufacturer's equipment – any D2B appliance will be able to communicate with any other D2B appliance.

So if you're a chronic worrier or have a memory like a sieve you'll be able to phone home and key in the code for 'Switch off the oven (or there'll be no supper and possibly no house by the time I get home)'. After fulfilling your instructions, the system will give you confirmation and you can start worrying about something else!

SMART HOUSE

●●○   ✓ CONVENIENCE   ✓ SAFETY

A few years ago it took a sizeable stretch of the imagination to envisage a VCR that could record your favourite TV programme while you were out. Today we take that for granted, and things have moved on. Now, just imagine a situation where all the appliances in your home can be operated from anywhere!

In the USA, the National Association of Home Builders, along with a consortium of manufacturers, research organizations and public utility companies, is working on the final stages of bringing this concept to the American public. The group has called its project the Smart House, and is preparing to launch the world's most all-inclusive system for home automation.

Once your home is wired, your wish is its command. It will turn the lights on for you as you come up the drive, and shut off the burglar alarm to allow you to enter without waking the whole street. You can flash every light in the house to let your family know that dinner is ready. Or you can instruct your oven to make the TV flash when the roast is done. Before you go to bed at night, a control panel will let you know the temperature of each room, and whether or not any appliances need to be switched off. And your stereo speakers, your phone, your television will work off any socket in your home regardless of where the stereo receiver, the phone point or the VCR are installed – all without additional wiring.

Everything will be controlled by a panel on the wall, a video touch screen, a remote control device, a voice recognition system, automatic sensors, and/or by telephone from anywhere in the world. Take your choice of any or all combinations – although voice recognition will cost you a bit more. Old appliances will work perfectly well with the system, but eventually appliances will be built with their own 'smarts'. Some will even be able to diagnose their own faults. A service engineer will call the machine on the phone to see what's wrong, or will come to the house and read the problem off the control panel.

But there's more to the idea than just convenience – Smart Houses will offer the ultimate in safety. If there is a gas leak, say, or any malfunction in the system, the energy flow will be cut down. If fire breaks out it will be visible on screens before it spreads. And when the plug of any electrical appliance is inserted into a socket, a microchip in the plug will tell the socket how much and what kind of current is needed. So if a child inserts a paper clip or a finger into the socket, no electricity will be activated.

'So what happens when there's a power cut?' the cynics will be saying. No problem. All that intricate programming won't be lost, because there's a back-up power unit that constantly recharges itself and will keep operative the basics of the system – clocks, instructions for timing, security system, and the pilot light for the gas. If you have set your teamaker to make you a cuppa at 7 a.m. and the power goes off in the middle of the night, the back-up can be counted on to produce the goods on schedule once the power comes back on.

MICROWAVE CLOTHES DRYER

●●● ✓ TIME-SAVING ✓ ENVIRONMENT-FRIENDLY
✓ MONEY-SAVING ✓ HEALTHIER

This is a new idea for the kitchen that seems to have a lot going for it. It will save energy and time, and may also kill germs while it dries your clothes. It's a space saver as well – a boon for people with small kitchens – because, unlike the standard tumble dryer, it doesn't require vents. So it can be tucked away in a cupboard, under the sink or under the stairs. And the time- and energy-saving claims? A small load would take only five minutes to dry!

Research has been going on at the University of Tulsa in Oklahoma, funded by a local company, Micro Dry Inc. According to Mary Ellen O'Connor of Micro Dry, the new machine should also be much kinder to fabrics than the conventional tumble dryer. 'In a regular dryer you heat the air, which heats the clothes. Then the water evaporates and you have to get rid of it. But microwave radiation eliminates one whole step of the process. You heat the water, not the air, directly. Clothes will come out much cooler.'

The Research and Development team in Tulsa has recently discovered that microwave drying also kills basic bacteria. So far it has been shown that seventeen different strains are disposed of in the drying process. Paul Kantor, president of Micro Dry, hopes that it will also be shown to kill staph infections, making the dryer useful in hospitals. He is particularly excited by the prospect that it may destroy viruses: 'Liquids are attractive to microwave energy and viruses are liquids. This could be a major breakthrough.'

SOLAR-POWERED COOKER

 ENVIRONMENT-FRIENDLY  MONEY-SAVING

Here's something else for the kitchen that will be equally at home on the patio. Forget about fiddling with charcoal briquettes and lighter fuel – now all you need is the sun to set those steaks sizzling on your barbecue! And it will work just as well if you use it indoors for the Sunday lunch. Alvin Marks, president of Advanced Research Development Inc., a Massachusetts company, has invented a versatile, solar-powered cooker called the SunCooker.

It works like this. An optical system that resembles a small satellite dish collects sunlight and sends it through a glass pipe into a cavity inside the oven. There, the energy is stored in what is known as phase-change material, which, like an ice cube in reverse, turns from a solid to a liquid as it stores heat, and resolidifies as it gives up heat. When the top of the cooker is raised, this stored heat comes from below, like a grill; with the top down, it works like an oven. The temperature is adjustable and can reach as high as 630° Fahrenheit (330° Centigrade), approximately frying temperature, for up to twenty-four hours before additional sunlight is required. For use indoors, the solar disc, which has sensors that let it automatically follow the direction of the sun, is placed on the roof. The light energy is then piped down to the cooker through long glass tubes like thick optical fibres.

The SunCooker couldn't have come along at a better time, as the need for cleaner, less expensive and more efficient energy sources becomes crucial. Indeed, Marks believes that use of his invention in developing countries could reduce deforestation by limiting the need for wood as a cooking fuel.

COOL COOKER

Here's a new idea in cookers that's even more revolutionary than using solar power. Toshiba's researchers have come up with a 'cool top' cooker that relies only on your pans to generate heat – but you'll have to throw out your glass and ceramic ones, because they won't work.

The principle is this: any metal utensil can become its own cooking ring when an electric current is induced in its base. The cooker top remains cool to the touch because your hand is non-metallic and an electric current can't therefore be induced in it. So if you spill food it doesn't get burnt on but can easily be wiped off.

To achieve this, ordinary domestic electricity is converted into a high-frequency current which is passed through a coil of conducting material. It generates a magnetic field which induces electrical currents in nearby metallic objects such as saucepans. Toshiba have had to develop a sophisticated coil that can adapt to the widely different electromagnetic properties of the various materials commonly used for metal pans – iron, stainless steel, copper and aluminium.

There's one current hiccup that might be slightly off-putting to some kitchen users. Under certain circumstances the coil's magnetic field could repel the current in a light pan to such a degree that it would hop about the hob and even levitate above it! Magicians, take note. Cooks, fear not – in this situation the control circuit will immediately cut the power.

There's also a possibility that strong magnetic fields may present a health hazard, and a considerable amount of further testing still has to be undergone. So it's probably going to be some time before the cool hob is a feature of our kitchens.

TOUCH-FREE TAPS

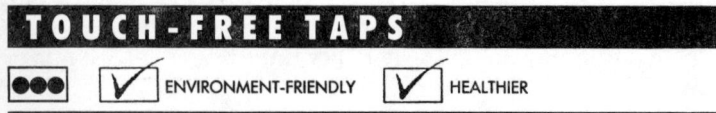

If you're concerned about hygiene or just like to conserve water, the touch-free turn-on tap pioneered by a Colorado firm, Mile High Precision Instrument Co., is something that you should know about.

The device, called Lampson's Automatic Mixing Faucet, uses infra-red technology and is connected to the domestic electricity system. When your hands or any other object are brought to within about 5 inches of the tap, an infra-red beam activates an electric valve that opens to let water out. The flow automatically stops when the object is moved away.

The main aim of the invention is to conserve water, a pressing concern at times of seasonal drought or where water is metered. Says Gordon Lewis, general manager of the Colorado company: 'One hundred and fifty families of four can save a million gallons of water each year using this faucet just to brush their teeth.' That's more than 6,000 gallons per family. The average teeth-cleaning session sends 3½ gallons of water down the plughole; with the new taps, only half a gallon or less is used.

The invention also appeals to current concerns about personal health and hygiene. 'You're not going to be as exposed to as many colds or transmittable diseases if you're not coming into physical contact with the faucet,' says Lewis.

INTELLIGENT LOO

●●○   ✓ HEALTHIER   ✓ WACKY

The self-flushing, self-sterilizing loo with a pre-heated seat is already delighting the Japanese with their own ingenuity. In the spirit of progress they are now developing an automatic loo-cum-bidet that obviates the need for toilet paper.

After you have completed your business a mechanical arm appears beneath you and shoots up a stream of warm water, followed by a blast of dry air that can gust for sixty seconds at a time. The full treatment is completed with a perfumed misting of your nether regions. Some of these automated geniuses even play gentle music!

One advertisement in Japan claims that to clean the bottoms of a family of four the toilet uses only half the electricity needed to run the refrigerator for the same length of time. Who could ask for anything more?

WALKING TV

A television on two legs is robot design at its wackiest, and that's what Brian Elliot intended when he invented the Animan. This TV of the future will be able to walk freely from room to room without human help. Elliot, a design student from California, has also equipped his prototype television with a security camera. When night falls, the Animan becomes the Aniscout as it patrols the home and sounds an alarm if an intruder is on the premises. Sony hopes to develop the invention and plans to add voice recognition as one of its features.

SOLAR AIR CONDITIONER

●●● ✓ ENVIRONMENT-FRIENDLY ✓ MONEY-SAVING

New Breeze is an air conditioner that uses the elements – water, air and sun – to keep a building cool. The brainchild of the Independent Power Company of North San Juan, California – where air conditioning is a necessity rather than a luxury. New Breeze is wall-mounted, will cool a 100 square foot room and requires no electric power. Instead, during the day it gets its energy from a 4 x 1 foot rooftop panel which harnesses the energy of the sun to generate the power that runs the unit. Inside the cooler is a fan which draws cold water from your water supply and then mists it out into the air stream. In the evening, if cooling is still needed, New Breeze can be run on regular AC current.

Although New Breeze currently costs as much to buy and install as conventional air conditioners, that initial outlay is easily recouped in the reduced running costs, says company spokesman Christopher Freitas. 'Owners can save at least $30 a month with New Breeze. They supply the water; nature does the rest.'

ROBOT LAWN MOWER

●●●

Hi ho, Silver! etc. Introducing the Lawn Ranger, a computer-guided robotic mower that behaves as though it has eyes and a brain. It's riding to the rescue of those summer weekend afternoons when you'd rather be on a golf course than on your own front lawn.

Developed by Technical Solutions Inc. of Virginia, the Lawn Ranger has sensors that let it 'see' where there's grass to be cut and the know-how to get there. First you have to guide the machine around the perimeter of your lawn by using a joystick on the hand-held control panel. After that, the Lawn Ranger takes over. It will steer itself around the garden, mowing any area where the grass exceeds the height you have pre-set for the sensors. When the job is done, the Lawn Ranger automatically turns itself off.

'What we have here is a computer that guides the mower, using sonar and sensors,' explains Ray Rafaels, the designer. 'The sonar helps the mower detect obstacles in its path, and the sensors detect the height of the grass.' Safety features include a bumper switch that automatically shuts off the mower if contact is made with another object, and a remote cut-off switch that allows you to stop the motor from a distance. The mower weighs 120 lb, stands about 2 feet off the ground, and is about 3 x 2 feet in area.

Rafaels and his partner, Jim Hammond, have a patent and a prototype and a lot of interest from companies who want to put the Lawn Ranger into production. Hi ho . . . and away!

FUTURE HOMES AND GARDENS

SELF-WEEDING LAWNS
●●● ✓ CONVENIENCE ✓ ENVIRONMENT-FRIENDLY

Grass that weeds itself: it's a home-owner's dream come true.

A geneticist in Canada has unearthed a grass that emits its own natural herbicide, an agent that destroys or inhibits plant growth. Not only is the grass self-weeding, it's lovely to look at. 'It makes for a beautiful lawn,' says Dr Jan Weijer, a scientist at the University of Alberta, who first discovered the grass in the eastern Rockies in the seventies while looking for grasses to battle erosion. 'The texture is rich and the colour is a very lush jade green,' he says, warning, however, that a dandelion may 'creep in every now and then'.

Because the grass originated in the dry, hostile soil of the Rockies, it is virtually maintenance-free. Apart from being self-weeding, it can grow in nearly any soil and climate, and it needs no fertilizer or watering. Also, 'the grass grows just four inches annually, so you only have to mow it about three times a year,' says Weijer. He is patenting the product in the United States. Marketing of the product should begin there in 1992.

NOISE CANCELLER
●●○ ✓ ENVIRONMENT-FRIENDLY

At last a quieter world will soon be ours courtesy of a new technology whose principle has been known to

31

scientists for years (RAF pilots have enjoyed it for the last decade), but was impossible to apply to the workaday world until recent advances in microprocessor technology opened the door. Now, everybody is coming up with applications.

A New York company called Noise Cancellation Technologies has developed the NCT 2000, an industrial noise and vibration control system that electronically analyzes noise, then matches it precisely with an anti-noise wave – one which is exactly 180 degrees out of phase with the original noise wave – and thus cancels it out, producing blissful silence.

The same principle is also being used by the Lotus car firm in Britain, who position microphones in the headrests; these are tuned to register the pressure and frequency of annoying noises from the engine. A microprocessor analyzes the signal and creates a counteracting effect.

Like Lotus, NCT is applying its system first to motor vehicles. Not only will it reduce engine and exhaust noise in cars and lorries, the company says, but it will relieve the vibration on the engine and therefore make it last longer. Other obvious pluses are more pleasant journeys and reduced driver fatigue – an important point for people who drive long distances, especially on motorways.

Lotus may be a luxury car manufacturer, but if you thought this refinement was going to be restricted to the rich, you can think again! It's often the cheaper models that are most in need of noise reduction, and since the technology isn't that expensive it's likely that they – and you – will be getting the benefit.

In America, after cars, NCT plans to apply the

technology to domestic appliances such as vacuum cleaners and lawn mowers. The one application which will have to wait is random noise in the home – the TV, children shouting, dogs barking. That's because the current technology can only gauge and match a consistent level of noise.

NOISE METER

●○○ ✓ CONVENIENCE ✓ SAFETY ✓ ENVIRONMENT-FRIENDLY

In case no one manages to work out how to cancel random noise electronically, here's a little device that should help you win arguments about heavy decibels thumping out from teenagers' bedrooms. For some time instruments have been in existence that measure general noise levels, but the human ear is the often unwilling recipient of 60 per cent more screeching, drilling, hammering and rock 'n' roll than whatever level is measured in an open environment. Now a small, portable, lightweight measuring instrument has been developed by British researchers for use at building sites and in factories.

The Personal Noise Dosemeter has a main unit that goes in your pocket or clips on to a belt. A small microphone permanently attached to the main unit is mounted with a small clip near the wearer's ear.

The amount of noise endured is displayed on a bright digital readout screen that can register from 0 to 99.9 per cent. At 100 per cent you have the maximum allowable exposure – pre-set by the user according to a chart. If you switch the instrument off, it will store for several hours the last 'noise dose' registered.

SOLAR LIGHTING SYSTEM

  ENVIRONMENT-FRIENDLY  MONEY-SAVING

Imagine growing flowers and vegetables in a room with no windows but with plenty of real sunlight! This seemingly impossible scenario is already a reality in Japan, where Dr Kei Mori, chairman of La Foret Engineering and Information Service Co. Ltd, has invented the Himawari (Japanese for sunflower) solar lighting system.

The main feature is a clear acrylic bubble, installed on the roof. It's a 6 foot wide solar collector that weighs 1320 lb (that's over half a ton, so you'd need a specially strengthened roof) and contains nineteen hexagonal lenses. Throughout the day these lenses follow the path of the sun, much the same way as a sunflower does. A solar cell communicates the sun's position to a microcomputer, which controls motors that point the collector in the right direction. When the sun is obscured by clouds, the computer signals a timer to reposition the collector when the sun reappears. At the end of the day, the computer turns the collector east for the next morning's sunrise.

Below each lens in the collector is a cable of tiny optical fibres, which deliver the concentrated sunlight to the rooms below. The light collected by one lens and sent through 130 feet of cable is equivalent to that of a 100 watt quartz halogen bulb.

An important feature of the system is that the acrylic bubble filters out most of the sun's harmful rays. Dr Mori claims not only that the filtered sunlight will grow plants, but also that it promotes the healing of various human ailments.

It's one of those inventions that's clearly going to be expensive to install – though very cheap to use. In the long run, it's a money-saver.

SOLAR ROOFING MATERIAL

  ENVIRONMENT-FRIENDLY  MONEY-SAVING

No, don't skip this as an 'old hat' idea – photovoltaic roofing should not be confused with solar heating. These are not panels installed on the roof to collect sunlight and convert it into heat. What we have here is a roofing material that is actually used in place of tiles. It keeps your roof from leaking at the same time as it provides free, clean electricity for the whole house.

Photovoltaic energy is a form of solar energy, but the key difference is that these layered semiconductor devices turn sunlight directly into electricity. Unlike solar water heaters, these solar cells have no moving parts, use no fuel except sunlight, and produce neither smoke nor noise. Sounds like a major breakthrough, doesn't it?

The principle of photovoltaic energy has in fact been

known about for some time: in 1969 it was used to charge batteries on a satellite in outer space. The difference is that, as with many exciting new inventions, with the passing of time the price has come down. While it was prohibitively expensive in 1969, the cost is now .01 per cent of what it was then – little more than the cost of ordinary roofing tiles.

Like most of these solar energy applications, this one is being developed where there's rather more sun than we see in Britain. Dr Heshmat Laaly, a California analytical chemist, is masterminding this wonder roofing. He's made the material flexible, so that it can be rolled up and transported to remote areas where electricity isn't available from the public system. The new roofing is easily installed and produces a minimum of 12 volts per square foot. The electricity runs down two wires into the house or other building, where it enters a converter. Any excess electricity generated during the day is stored in the equivalent of eight or so car batteries.

Dr Laaly is looking for a few 'sincere millionaires' to get his product off the ground. (Note to sincere millionaires: Dr Laaly's invention won the Most Commercial Potential Award at the 1988 International Invention Convention.)

The windows in your home through which you see out and the sun sees in will some day also capture the sun's energy and convert it to electrical power.

Transparent amorphous solar cells have been developed by researchers at Sanyo Electric in Osaka, Japan. They collect solar power in the same way as the bulky solar roof panels with which we are familiar. The difference – and it's a substantial one – is that these new cells have been constructed using an exclusive technology that etches delicate microscopic holes throughout each cell, which render them transparent. Natural light can pass through the window cells to light up your living room while, simultaneously, the rays are being tapped to run your television set.

Sanyo plans to market transparent solar cells for skylights, car sun roofs, greenhouses and conservatories, as well as for windows in the home. Another possibility on the skyline is that the cells may be used as a power source for yachts.

PRIVACY WINDOWS

✓ CONVENIENCE ✓ SECURITY

Curtains and blinds could soon be a thing of the past if a new product catches on. Called Varilite Vision Panels, they work like this. Sandwiched between two panes of glass is a liquid crystal layer, similar to the LCD displays in digital watches. The glass is smoky until a switch is flicked on, when the glass becomes clear. The system is run by electricity and controlled by any standard lighting switch, timer or computer.

Manufactured by the Taliq Corporation of Sunnyvale,

California, Varilite windows are already being used in some conference rooms, limousines and hospitals and should be available to us all well before 2001. They'll even come in different colours: clear, bronze and grey.

WINDOW SHATTER PROTECTOR

●●● ✓ SAFETY ✓ SECURITY

Here's something that could be literally life-saving. A group of Dutch researchers has perfected a transparent covering that allows a window to withstand the blast of a hand grenade without shattering. The product, called Profilon Plus, is a transparent sheet made of polyester fibres that sticks to the inside of ordinary glass windows.

During World War II, the scientists recalled, people would criss-cross their windows with tape to prevent them shattering during air raids. So the Dutch team took that simple idea and improved it considerably. Profilon Plus is a lot tougher than tape – it is claimed to increase the strength of glass by 300 per cent, and an impressive promotional video shows a stick of dynamite exploding while fastened to a window: the blast fails to shatter the glass.

And in parts of America and other countries which suffer more extreme weather than Europe there's an additional incentive to fitting your windows with this product: it offers protection against flying glass during hurricanes and tornadoes.

FIRE EMERGENCY LIFELINE

SAFETY

The vast majority of deaths from fire are caused not by burning but by smoke inhalation. When Ray Tannatta, a New York fireman, found one such victim collapsed in the bathroom of a burning block of flats he realized that an unlimited supply of breathable air lay just inches away in the drainpipe of the washbasin. As a result he invented the Lifeline, a breathing device that consists of two plastic air masks connected to a 7 foot hose. If fire breaks out this hose can be hooked up to a sink, bath or basin drainpipe, allowing two people to obtain air through the pipe while waiting to be rescued.

And in case you're worried about whether drainpipe air is healthy to breathe, Lifeline Services Ltd says it is. The product comes equipped with an activated carbon filter to prevent inhalation of any noxious fumes. There's a home model, of course, and also, very sensibly, a travelling one which will be sold in hotels and airport shops and packs into a suitcase.

MEMORY METALS

SAFETY

'Metals with a mind'? Sounds like sci-fi of fifty years ago? In fact it's just a snappy phrase to describe a discovery that could save lives and make fire-fighting an easier job.

An alloy made of nickel and titanium (known as SMA

for shape memory alloy) has an almost unique ability to 'remember' a certain shape and go back to it when it is heated to a pre-determined temperature. Other metals share this ability, but are unfortunately susceptible to stress or corrosion.

There are many applications for this extraordinary substance. Among the most important are self-opening hatches that could make access to roofs easier when buildings are on fire; similarly, SMA could be used in emergency fastenings to close public litter bins if vandals set fire to them. Dentists envisage memory metals being used for braces that will adapt their shape automatically as children's teeth grow. And a Japanese electric percolator using an SMA coil is said to make excellent coffee! Surely many other uses will suggest themselves to researchers as time goes on.

ELECTRICAL SHOCK HAZARD PROTECTOR

SAFETY

This is a little device with a big name – Immersion Detection Circuit Interrupter (IDCI). Its function is to disconnect the power when water is present in the current. In other words, if your hair dryer or radio falls into the bath with you (not that you should be using them that close to the bath, of course!), you should be protected from electric shock or electrocution.

The device itself, manufactured by Leviton of New

York State, won't be available to the public. You will have to buy a product ready-equipped with an IDCI.

VOICE KEY
●●● ☑ CONVENIENCE ☑ SECURITY

No more fumbling in your handbag or pockets for your house keys. In the future, the mere dulcet tones of your own voice will open doors for you. That's the idea behind a Massachusetts company's new security system – VoiceKey.

Whether you hail from the industrial north or the farmland of the West Country, whether you stutter, lisp, mumble or carefully modulate your words, it matters not to VoiceKey, a device that uses the individual patterns of each human voice as its combination for unlocking doors.

In this latest innovation in voice-verification technology, it takes the user only a single twenty-second session to be 'enrolled' into VoiceKey. A voice sample is spoken into a built-in microphone and then stored in the compact unit that is mounted on the wall next to your door. The user also chooses a personal identification number (PIN) and the password, which is spoken. To gain entry past the area that VoiceKey guards, a user punches in his PIN and utters the password. In half a second, a decision will be made by VoiceKey to grant or refuse access. Each time access is given, the system updates the acoustic voice pattern of the user to reflect daily fluctuations.

Michael T. Dougherty, sales manager for the manufacturing company, explains that what makes a system such as this more effective than keys or cards is that, like fingerprints, no two voices are ever exactly alike. And while the voice reader allows for natural changes in the human voice – croaky, hung-over, streaming with cold – it will not respond to recordings.

HAND-SCANNING LOCK

●●○ ✓ CONVENIENCE   ✓ SECURITY

Here's a product, though not yet advanced from the drawing board, that relies on the uniqueness of certain human attributes for its effectiveness. This one will read your handprint with a digital scanner before deciding whether or not it will unlock the door for you.

It is being developed for high-security offices and for banks, which can apply the scanning device to their automatic machines. Instead of punching in your personal number, you'll just place your palm on the machine's screen. The computer will then compare your hand to a digital description of your own unique handprint.

Applying the same concept to house locks will be the next step for this technology, says Tom Kelley of the California company that is creating the model scanner for the manufacturer, Biometrics Inc. 'You could have your front door programmed so that only certain people could come right in,' he says.

The prototype hand scanner is a rectangular box with

a red plate on top, and an outline showing the proper position for the palm. Smaller than a cigar box, the mechanism could be mounted into the wall of a porch or near a garage door.

Like many inventions, it will probably be expensive to start with, says Tom Kelley – perhaps two or three times the cost of a combination door lock. For this reason it is likely to appear first on prestige private estates and similar housing.

ROBOT DOG

●●● ☑ WACKY ☑ CONVENIENCE ☑ SECURITY

Whether this Australian idea will ever have quite the same security potential as the VoiceKey or the Hand-scanning lock is doubtful – but it's probably more fun!

This 'pet' can obey fifteen commands, turn in different directions, pick up objects and detect a human presence. And it has an infra-red sensing system so that it doesn't bump into furniture or walls. The dog doesn't bark, but it can speak. It can issue a warning message to intruders, and its synthetic speaking voice also enables it to ask and answer questions.

Inventor William Holden came up with the idea at a robot show in the USA. 'I tried to combine the idea of technology with the concept of a pet,' he says. Back in Australia, he watched how the koala bears moved around at the zoo. The robot dog was developed with those observations in mind.

This version of man's best friend runs on motors powered by rechargeable batteries. He does not, of course, cover the furniture with his hair; nor does he have to be taken for walks or leave little batteries on your carpet.

NON-CHOKING DOG COLLAR

●●● ✓ HEALTHIER

If you're a traditionalist who likes his or her dog to come in recognizable hairy form – the price of dogfood and vets' bills and the inconvenience of those late-night walks in the rain notwithstanding – here's a little thing to make the life of man's best friend, a little more pleasant.

'We train horses, cattle, sheep, goats and calves with halters, yet it is accepted practice to train dogs by choking,' says Dr Robert Anderson of the Animal Behavior Clinic at the University of Minnesota. He and a colleague have therefore come up with the Gentle Leader, a dog 'halter' which takes into consideration a dog's natural instincts.

Gentle Leader consists of one strap that fits round the dog's nose and lower jaw, and another that runs behind the ears and joins under the jaw. 'By controlling the animal's head and nose,' Dr Anderson explains, 'you are better able to control the whole dog.' The halter-type collar removes the pressure from the dog's throat and applies it to the nose and large muscles at the back of the neck. 'Female dogs take their puppies by the scruff of the neck to control them,' observes Dr Anderson. 'Older dogs remember this and recognize pressure to the back of the neck as control, and respond accordingly.' When you pull and tighten the nose strap, the dog's mouth is forced closed, preventing barking or biting. But unlike the conventional muzzle, which can cause a dog to panic, the strap loosens as soon as you slacken the lead. Rover soon learns that with good behaviour he's going to be more comfortable.

With Gentle Leader anyone, no matter how weak, can control any dog, no matter how big, and inflict no pain on the animal. It's Dr Anderson's hope that his invention will reduce the number of dogs that are given away or sold, or put to sleep, just because their owners regard them as untrainable.

'BARK STOPPER' DOG COLLAR

✓ CONVENIENCE

Here's something for Rover that relies on his undoubted intelligence to work out what makes life easiest for him. This, too, has been dreamed up by veterinary researchers at an American university.

If your dog is considered the local nuisance – making you unpopular and your pet a prime target for a hit-and-run accident – your troubles are over. This special collar looks very similar to the standard sort, but has a small speaker attached. When the dog begins to bark, the collar responds by emitting a high-frequency sound that is inaudible to humans, yet unpleasant to the dog. Once he's heard that sound a few times he begins to realize that it only occurs when he barks. As a result he stops barking and you stop receiving calls from angry neighbours. Even chronic barkers can be cured within a few hours, say the inventors.

But fear not – the collar will not turn your dog into a wimp! If he is barking because a burglar has broken into your home or because you are being threatened by a mugger, his natural instincts will take precedence over fear of the collar. Loyalty will overcome discomfort.

BUTLER-IN-A-BOX

✓ WACKY   ✓ CONVENIENCE   ✓ SECURITY

'I say, Jeeves, what d'you think Aunt Agatha would say to this jolly old invention?' . . .

Butler-in-a-Box is a wonderfully eccentric enhancement to automated house technology, and rather more sophisticated than the Robot Dog. Gus Searcy, a professional magician and computer buff, got tired of people asking, 'Can you make the lights go on?' whenever he pulled rabbits out of hats. So he designed this voice-activated box out of which anyone can pull a butler, and now it's manufactured by Mastervoice Inc. of California.

A mini-Smart House, it will not only turn on the heat in the upstairs bedroom, or turn off the light in the bathroom, but will perform any of thirty-two commands you give it. Using voice, timed control and keypad, the

Butler can control a total of two hundred and fifty-six household devices.

But first the 3 x 9 x 11 inch box needs to be given a name – let's call it Jeeves – because it thinks it is a real butler, and it needs to be trained. A few entries are made on its built-in keypad, and then you must recite ten phrases so that Jeeves knows what you sound like. Finally you must enter each desired command that you will want Jeeves to perform. Jeeves can be programmed to be charming or cantankerous, as you wish, and can respond in a foreign accent or a foreign language.

Ask him to book you a table at your favourite French restaurant, and Jeeves can find the correct number in his files and dial it for you. He'll also activate the telephone if a call comes in; you can talk over the speaker phone to the caller. And, like his human counterparts, if Jeeves makes a mistake he apologizes.

Jeeves can pick out his master's voice in a crowded room and won't be confused by the television. He is trained to respond only to the four voices entered into his system. If a burglar breaks in, he will ask, 'Hello, may I help you?' Without the correct response of a secret password, Jeeves will throw a veritable tantrum: all the lights in the house will start flashing, the stereo will blare and the burglar alarm will go off.

SMOKE CHECK BADGE

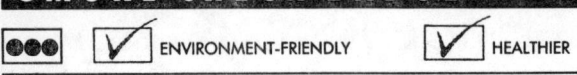

✓ ENVIRONMENT-FRIENDLY   ✓ HEALTHIER

With non-smoking now a burning issue, here's a little device that should settle a few arguments. If you don't smoke and your partner does, you'll be able to tell just how polluted your home is. If you work in a non-smoking office but colleagues in the vicinity are puffing away, you'll be able to discover just how much smoke is wafting over to your desk and causing unwanted passive smoking.

The badge is a disposable device that turns from pale yellow to deeper shades of brown as it registers accumulated exposure to tobacco smoke in the air. It has been developed by Assay Technology, a California company that manufactures industrial products to measure toxic chemical levels in the workplace.

What Smoke Check can't do is give you a reading in minutes or even hours, so it won't settle that argument in a restaurant between a smoker and a non-smoker. Since it takes three to five days for an accurate reading on low-level exposure, its use is really limited to places you spend a long time in – the workplace and home.

DOME HOMES

✓ ENVIRONMENT-FRIENDLY

A man's home may be his castle, but a man's dome is his energy saver. With worldwide focus on conservation – ecological and monetary – will villages of dome homes be dotting the landscape of the future?

The geodesic dome-shaped home is not for everyone, concedes Michael Busick, president of the Florida company responsible, American Ingenuity Inc. If you enjoy the material aspects of life then a dome home, no matter how efficient, probably isn't for you. If, on the other hand, you're concerned about the environment and what kind of world your children and grandchildren will inherit, then you might possibly support Busick and his wife's belief that the dome is the home of the future.

First and foremost is the matter of energy. The Busicks have patented a non-toxic insulation process that, they say, makes their dome homes three to four times more energy-efficient. Instead of using fibreglass American Ingenuity uses polystyrene foam triangles, which are chemically stable, will not rot or deteriorate, and are non-toxic even if burnt.

Polystyrene is not often used in the construction of traditional homes because it is more expensive than fibreglass. But because the surface of a geodesic dome is smaller than that of a square-shaped house (if you don't know what a geodesic dome looks like, imagine a golf ball with the little indentations turned into fewer and larger facets), less polystyrene is needed. The combination of polystyrene insulation and concrete exterior keeps homes warmer in winter and cooler in summer than their traditional counterparts.

The dome is as efficient in colder climates as it is in warm ones. Though it is clearly viable whether we shall be seeing them in Britain remains to be seen. America boasts some sixty of these homes, which are available over there in kit form.

2
TWENTY-FIRST CENTURY COMMUNICATIONS

Pocket computer
Pocket organizer
Computer alarmcard
Write-top computer
The ultimate computer notepad
World's Smallest Weather Station
Weather cube
Water battery
Solar-powered briefcase
Walking Desk
CD ROM
Large-capacity smart cards
Video telephone
Holographic phone
Selection telephone
Telephone voice changer
Voice-activated typewriter
Portable voice-activated translator
Watch pager
Prayer wristwatch
Panic alarm wristwatch

POCKET COMPUTER

☑ ENHANCEMENT ☑ CONVENIENCE

From calculators to television sets, everything electronic is going pocket size. But how do you miniaturize a computer screen without losing most of the image or making the text so tiny that it's unreadable?

Picture a home computer with a full display of graphics and text up on screen. Now lose the computer terminal and the keyboard. Picture only that full image of graphics and text floating by itself a couple of feet in front of you. Now come back to reality and imagine the physical screen shrunk to half the size of a cigarette packet.

And there you have the solution to the pocket computer problem. The small screen has a 1 inch square window. When you look into it, you will see the equivalent of what you would see on a full 12 inch screen; just the information will appear to be floating in space a couple of feet in front of your eyes. The mini-screen could be clipped on to a pair of glasses, or hang down over your eye from a headset. The whole operation will work much the same as a Walkman: the visual display screen will be wired to a mini-computer – with a keyboard – that will be carried in your pocket.

Steve Lipsey, who spends his time explaining this technology to electronics companies who will eventually be using it, is vice-president of Reflection Technology. This Massachusetts company is marketing the visual part of the pocket computer under the name Private Eye. 'We've freed up the electronics industry to make their products in a very different way,' says Lipsey. 'They no longer have to restrict themselves to putting their products in boxes large enough to read.'

Some day the technology will be used for all sorts of applications, from miniaturized fax machines to pagers that can receive road maps. And surgeons will use the Private Eye to check on a patient's vital signs while on the operating table.

POCKET ORGANIZER

●●● ✓ CONVENIENCE

Filofaxes are as passé as last year's Porsche. So what is new on the designer information retrieval scene?

The AgendA Microwriter slips into your pocket and won't make you feel you are carting a telephone directory around in your pocket. It has all the features you would expect of an organizer, but much more: a superb filing system; it can talk to PCs and be used interactively with a Mackintosh; it has German and French language packs; and an excellent wordprocessor with some nice tabulation features.

The British, award-winning AgendA can be used in the hunt and peck mode, but it also has a set of special microwriter keys, which are pressed in combination for touch typing. The alphabet takes half an hour to learn and touch typing at thirty words per minute is soon attainable for the average user. The keyboard is silent, and requires very little effort or movement of the fingers.

For those writers who get their best ideas when they

are lazing on the beach, and who forget them on the way home, the AgendA is a must.

COMPUTER ALARMCARD
SECURITY

To hear inventor D'Arcy Dawe talk about the computer industry, you would never guess he's in the insurance business. Of course, if his computer alarm does as well on the market as he hopes, he may not stay in the insurance business for long.

Dawe's product is the Alarmcard, a 4 x 6 inch motion-sensitive board that plugs into personal computers and sounds a loud alarm if the machine is moved. Running on a 9 volt battery, that Alarmcard plugs snugly into the PC's interior, where it can't be seen by would-be thieves.

Dawe would like to see the Alarmcard become a standard feature in personal computers. 'My immediate goal', he says, 'is to lobby insurance companies in order to gain support for security items in computers.' However, Dawe has higher hopes for the technology behind Alarmcard than computer security alone. He hopes to develop his baby further, so that it can be reduced to a chip that would thief-proof any device containing a microprocessor. Any item that can be programmed could then be wired easily and cheaply to sound an alarm if tampered with.

The Alarmcard can conveniently be monitored by a central security post, making it a worthwhile investment for larger companies, educational establishments, government installations and so on.

WRITE-TOP COMPUTER

●●● ✓ CONVENIENCE

It may look like a toy – but that's where the likeness ends (though it is easy to operate). This is a portable PC that actually lets the user input just by writing. Several versions are currently available, including Write Top from Linus Technologies Inc. of Virginia, and Penpad, made by Pencept of Massachusetts and marketed in the UK by Bergman and Co. of London.

'The great thing about the Write Top is that you don't have to know how to use a PC. You only have to know how to write,' says Brooks Puckett of Linus Technologies. The user makes notes, then pushes a button to transform his handwriting into print. You have a chance to edit for mistakes – the computer may have misread something – before saving it on disc. Mistakes are eliminated simply by writing over notes or by drawing a line through an entire sentence.

These clever little machines cut out the laborious step of inputting into a computer notes you have already taken by hand. And it reduces the risk of error. 'The inputter may mis-key,' says Puckett. 'Or, if you're using a secretary, she may misread your handwriting. Then you've got the added chore of going back and proofreading everything that's been keyed in.'

The Write Top, which in all other respects can serve

as a standard IBM-compatible PC, has various extra features. A glossary option allows the user to call up frequently used information – phone numbers, client contacts and so on – at the push of a button. And the computer can be used for filling out forms, which are drawn on to the screen and saved for repeated use.

So why haven't we had this useful little gadget before?

The problem that the computer industry has been trying to overcome for years is character recognition, because people's handwriting differs so greatly. And while a computer can be programmed to read writing in a particular style, this too is impractical because individuals are not consistent. To get over this hurdle Pencept developed a piece of software which recognizes characters by decoding the pen strokes from which they are composed. This, apparently, is easier than trying to register overall shapes of letters and numbers.

Pencept appreciates that many people can type as fast as they can write, and with less fatigue – so what's the use of this product? The company sees the Penpad tablet coming into its own where people are doing several tasks simultaneously – such as putting text in via a keyboard and entering graphics on a tablet. Because the machine can recognize handwriting the whole lot can be done on the Penpad without switching around.

THE ULTIMATE COMPUTER NOTEPAD

✓ CONVENIENCE ✓ TIME-SAVING

This is the computer of the future. It will do an enormous range of things, yet be light enough to carry around. This dream machine is the brainchild of five students and two lecturers at the University of Illinois, who won a competition sponsored by the Apple Computer Co. to come up with a personal computer for the year 2000.

Their invention is called the Tablet. There's no prototype yet, but most of the technology is here and the rest is on the drawing board. Bartlett Mel, one of the students involved, explains that the Tablet is nothing like the present-day PCs – a box with a keyboard. 'Our machine is completely portable. It's the size of a standard notebook pad and weighs one to two pounds.' There are no buttons, knobs or keys. Now here's what the Tablet will enable you to do:

1. Using a no-ink stylus, you'll be able to write and receive messages.

2. You'll be able to watch television on it. (Or, to be more precise, you'll be able to watch sixteen squares relating to sixteen different channels simultaneously.)

3. Around the edge of the Tablet will be infra-red sensors – like the ones in your TV remote control gadget. These will enable the Tablet to talk to other Tablets in the same room, or to any other compatible electronic equipment – including full-size computers.

4. The Tablet will come with an optional wall-size screen so that you can enlarge anything you display on

the small screen.

5. You'll be able to insert optical laser cards – the floppy discs of the future – which are the size of credit cards. Each will be able to store four hours of video or two thousand books. So you'll be able to read or watch pre-recorded videos on your handy computer.

6. The Tablet will double as a cellular phone. Touch the phone icon, give the computer the number, and you can communicate with anyone anywhere in the world. Your voice will be transmitted by a built-in microphone. And using the optional lapel-size camera accessory, you'll even be able to send video images.

If you think that's enough, there's one more bit of magic. Using the US government's Global Positioning System, Tablet will tell you exactly where you are and give directions to any place you want to be.

And should you lose your precious computer, Tablet can phone home and tell you where to find it!

WORLD'S SMALLEST WEATHER STATION

●●● ✓ CONVENIENCE

With the World's Smallest Weather Station, you can have the latest forecast in the palm of your hand. Weather enthusiasts, sailors and budding meteorologists can use this battery-operated, computerized gadget to measure wind speed, wind direction and air temperature.

The wind assembly mounts easily on a home antenna or the mast of a sailing dinghy, and transmits data to a hand-held station. Its computerized memory allows the user to record readings over an extended period of time. And for a little extra outlay you can add the option of a rainfall monitor, which measures rainfall to within a tenth of an inch.

The World's Smallest Weather Station is at present available by mail order in the USA. Who knows – if it had been available in Britain a while back we might have been better forewarned about the Hurricane of '87!

WEATHER CUBE

|●○○| ✓ CONVENIENCE

What's the weather doing? In future we won't have to check the newspapers, catch the TV weather report or even look out of the window. We'll simply glance at our weather cubes.

Currently the hottest gadget in Japan, Toshiba's 4 inch cube is a battery-powered device that predicts the weather eight hours in advance. The LCD screen displays the forecast with an appropriate symbol – sun, clouds, rain or snow. For the Japanese it's now an essential travel accessory – at picnics, it warns of rain; on the ski slopes, it hails the next snowstorm.

The device is actually a microcomputer with a program that's based on forty years of weather patterns. Unlike a barometer, with its old-fashioned mechanical system dependent on air pressure and humidity readings, the weather cube uses a semiconductor for readings of current weather conditions and comparisons with past weather data.

Japan's small size allows it to have relatively uniform weather conditions. So although it's been found impossible for that reason to have a single weather cube for the whole of the United States, the cube should work ideally in Britain when it will be marketed over here.

WATER BATTERY

●●● ✓ ENVIRONMENT-FRIENDLY ✓ MONEY-SAVING

Since so many of these sophisticated gadgets actually need old-fashioned batteries to operate some part of them, here's a piece of good news: a battery that runs on water, or orange juice, or Coke – or even a Bloody Mary! Sounds too bizarre to be true? But a man called Roger Hummel says he has one, and that it can last up to fifty years.

Hummel is director of VentuResearch, a small Texas firm that specializes in product development; its clients have included IBM, ITT, AT&T, General Motors and General Electric. A few years back the company came up with a liquid-powered watch that enjoyed moderate success; the water battery is an extension of that technology.

Conventional batteries run on a chemical mix and last anything up to three years, depending on size. They 'die' when the chemicals corrode the battery's innards. Also, standard batteries start to discharge from the moment they're manufactured, wearing down even as they sit on shop shelves. 'Our battery presents none of these problems,' claims Hummel, 'because it runs on liquids instead of toxic chemicals.'

To use a water battery you just unscrew the cap, pour in the liquid of your choice, screw the cap back on, and away you go. When you've finished, you just rinse out the battery and store it dry. The inside won't corrode because the 'juice' is drained, and the battery doesn't die because it discharges only when there's liquid inside. Now no one would want to empty a heavy car battery every time they used the vehicle, but with such ingenious minds as these surely only time will be needed before this and any other problems will be overcome.

SOLAR-POWERED BRIEFCASE

●●● ☑ CONVENIENCE ☑ MONEY-SAVING

While we're talking portable, here's a useful gadget which will be here long before the Tablet. The high-tech briefcase of tomorrow won't be carrying battery-powered (water or otherwise) pocket calculators, dictating machines and the like. Instead it will house a wide variety of business tools, all powered by solar energy. A solar panel on the side of the briefcase produces electricity, which charges a battery pack that will run the portable equipment within. Samsonite is already making the cases 'on special order', but in the future solar-powered briefcases should be every commuter's standard accessory.

You'll be able either to buy the equipment separately, or to purchase a custom-designed briefcase complete with equipment for your special needs. A possible package might include a lap-top computer, a cellular phone and a fax machine. And there would be specialized ones to include medical equipment for doctors, for instance.

As with a computer, the more you want in terms of add-ons, the more you'll pay. At the moment prices are high, but they are bound to drop dramatically when solar-powered briefcases are mass-produced.

WALKING DESK

✓ HEALTHIER ALTERNATIVE

Once you've got to the office (or even if you work from home as we will increasingly do in the future), sitting at a desk for long periods gets everyone down. So Nathan Edelson, who runs an American company called Environments for Health, has come up with the Walking Desk to deal with what he calls 'postural fixity' – a condition that may cause aches, pains, fatigue, stress and eventual illness.

His idea consists of a computer work station that stands higher than a normal desk and has a treadmill, exercise bike and stair climber installed underneath. On a good day, an office worker should be able to walk four or five miles and burn as many as fifteen hundred calories while dealing with his or her normal workload.

Apart from these exercise features the Walking Desk boasts a compact disc player for listening to calming music, and a colour monitor for viewing relaxing scenes of nature. 'When you have a client screaming at you on the phone,' reasons Edelson, 'turning up the speed on the treadmill and gazing at scenes of foaming surf on the "video window" makes a lot more sense than returning your caller's hostility.'

Weight loss, better muscle tone and reduction of stress seem like a pretty good package to offer. But there's more than that. 'Best of all,' claims Edelson, 'people say it makes them feel better, and that it's fun to walk at work.'

CD ROM

✓ CONVENIENCE ✓ TIME-SAVING ✓ SPACE-SAVING

Research can be a drag: you end up spending more time locating than actually learning anything. But if looking things up is a big part of your life – if you're a student, a journalist or professional researcher, say – here's some happy news. Called CD ROM, it's a 5 inch silvery circle that stands to revolutionize libraries. To use it you'll need a special CD drive unit that plugs into your current computer.

CD ROM stands for Compact Disc Read-Only Memory. These discs are similar to those used by audio enthusiasts except that, instead of music, they contain text and graphics like a book. And just one CD ROM holds as much information as 1500 floppy discs of the kind now used in computers.

Some software companies have packed complete stacks of fat reference tomes on to a single disc; whole encyclopaedias and the twelve-volume Oxford English Dictionary, with 252,000 main entries, are available on CD

ROM already. In the USA, the full text of every patent filed since 1971 can be searched out via CD ROM and displayed on screen within twelve seconds. The British Library is experimenting with a similar scheme.

A few discs with combinations of various volumes could put a library at your fingertips. It's every researcher's dream come true – as it is for those computer pioneers who envisaged volumes of information pouring across small screens around the world.

LARGE-CAPACITY SMART CARDS

☑ CONVENIENCE   ☑ TIME-SAVING   ☑ SPACE-SAVING

This is what the real future in information storage looks like. Large-capacity smart cards will eventually replace floppy discs because they are two hundred times faster at retrieving information yet small enough to fit into a wallet or purse. They're more versatile than CD ROM, because you can erase and input information. And although they'll cost you more than a floppy disc, there's probably an overall saving when you consider their vastly greater storage capacity.

How much greater? On a space the size of an average credit card you can store two thousand pages' worth of information. The best of the floppies can only do just over a quarter of that. And all this information can be retrieved on a reader as fast as you can flip channels from BBC to ITV.

The floppy disc will compete nobly with large-capacity smart cards as engineers improve the discs' storage capacity. But you'll need a personal computer to handle them (rather more expensive than a smart card reader). Laser optical cards will also be up there in the running, but they'll be used primarily in industry because their readers cost so much. And again, CD ROM will be for reading only. Smart cards – small, fast, with high capacity and low-cost readers – could well be the first choice of consumers in the nineties.

VIDEO TELEPHONE

●●○ ☑ ENHANCEMENT ☑ CONVENIENCE ☑ MONEY-SAVING
☑ SECURITY ☑ TIME-SAVING

Twenty-five years ago a 'picture phone' was shown at the New York World's Fair. Now a video phone is about to arrive on the telecommunications scene. The UV Communicator may well be the first marketable video phone of its kind; developed by a Californian company, UVC Corporation, the phone should lead the way for enhanced telecommunications between households, offices and businesses.

The system comprises a colour video camera; a 13 inch colour video monitor; a transceiver (an apparatus that alternately transmits and receives) with software for recording and transmitting video images over ordinary telephone lines; and a telephone for transmission of audio signals over a second telephone line. The audio quality is similar to that of a standard telephone, while the video resolution approaches that of a regular TV screen.

The video phone also comes equipped with an

electronic 'blackboard', through which images drawn on the screen with a light pen can be sent to a recipient's screen. It can also capture and store permanent records of visual images on a 3 inch floppy disc at the touch of a button. The phone transmits only the part of the picture that moves. Background images remain static, and software irons out any jerkiness.

'We are convinced', says John Looney, president of UVC, 'that in time every home and office that currently uses a telephone will one day use the video phone.' He's not boasting in vain. Since the video phone can be linked to ordinary phone lines the product has virtually unlimited potential for domestic, commercial and industrial users alike.

Looney foresees the Communicator particularly benefiting areas such as finance, property and medicine. The time and expense of out-of-town appointments can be reduced by face-to-face meetings over the phone; estate agents can show property to residential or commercial customers without unnecessary travel; and doctors can consult around the globe about their patients, enabling life-saving decisions to be made in seconds. Families split and spread around the world – through emigration, or working abroad – can 'welcome their first grandchild, be part of the reunion they couldn't work into their vacation, or watch the baby take her first steps,' Looney explains.

HOLOGRAPHIC PHONE

☒ ENHANCEMENT ☒ CONVENIENCE ☒ MONEY-SAVING
☒ TIME-SAVING

Here's an idea that uses a more modern technology – holography. Imagine looking into a 4 inch cube at a three-dimensional image of the person you are speaking to on the phone. Sounds like an end-of-the-pier What the Butler Saw machine? Professor Stephen Benton, who teaches physics, computer science and electrical engineering at the Massachusetts Institute of Technology, says such an invention is not too far off.

Benton's phone would use a holographic image – and he is renowned for having developed the best of these 3-D images. It would make use of a camera with two lenses, installed in your home, and would need to be hooked up by a fibre optic line to a super-computer, presumably operated by the phone company. The super-computer would take the images of both camera lenses, analyze them, and transmit a fully holographic image to the person on the other end of your line.

The problem with this technology is that the phone needs a 'dedicated' fibre optic, meaning that no other use can be permitted on the line. It is expected that in the future every home will have a fibre optic line, but the 3-D phone would require households to have two lines if they wanted any service apart from this one.

We aren't likely to be seeing this new development until well on in the decade – Professor Benton says that holography seems to be on a commercial back burner at present. 'The scientific and technical applications are driving this now, not use by consumers, so it's not getting a whole lot of attention.' And when asked about holographic TV, Professor Benton reports that it would take several fibre optic channels to get into your home and could be twenty years away. So for now he is thinking only about the 3-D phone when he says, 'I hope it happens!'

SELECTION TELEPHONE

●●○   ✓ CONVENIENCE   ✓ SECURITY

For those who, like Greta Garbo, desperately 'vant to be alone', Sanyo's Incoming Calls Selection Telephone is just the ticket – and much less expensive than employing a secretary to screen your calls. It's also different from an answering machine in that it won't take messages – but then it won't lose them, either!

When callers dial your number, they will hear three rings (though you won't hear a thing), after which a pre-recorded voice will come on the line and ask for a four-figure, pre-determined secret code that you have assigned. If these numbers have been registered with your machine beforehand, the call will be accepted and the numbers will be displayed so you can tell who is calling. But if the caller hasn't been given a code number by you, he or she won't be able to get through. The call will automatically be disconnected unless an acceptable code is offered within fifteen seconds of the third ring.

The device, invented by Sanyo, has an on/off switch

so you don't have to use it all the time. With the ability to store up to thirty codes, it is equipped with an automatic dial that stores up to ten outgoing numbers, and has on-hook dialling. It even boasts line-hold Muzak-like versions of 'Hey Jude' and 'Yesterday'.

At present the Selection Telephone is available only in Japan, and Sanyo say they have no plans to market it in other countries yet. But like most Japanese innovations it's sure to get here in the end.

TELEPHONE VOICE CHANGER

✓ SECURITY

Here's a form of telephone security that may appeal to you. The digital voice changer can make a man's voice sound like a woman's, a woman's sound like a man's, or any voice sound like another.

So who – apart from the elderly and/or female who might be bothered with heavy breathers and the like – would want to do any of these things? Well, if you're a spy, a private detective, or a husband or wife who's having an affair, a change of voice might come in handy. To help things along, the digital voice changer also comes with background noises, such as a barking dog (to discourage would-be harassers) or building site sounds (to give the impression you're somewhere else).

If all this sounds a bit James Bondish – well, it is. The voice changer is made by CCS Communications Control of New York State, who manufacture a full range of security devices. Called the DVC-1000, it comes in two compact pieces. There's a hand-held microphone and master control with a dial to adjust your voice from very low to very high, and another dial for background sounds. Attached by a cord is a cup-like device that fits over the mouthpiece of your telephone. At the moment it's only available in a few cities in the USA, but it should soon be much more widely distributed.

VOICE-ACTIVATED TYPEWRITER

✓ CONVENIENCE ✓ TIME-SAVING

For office workers, students, journalists – anyone who strains over a typewriter or word processor – this idea seems too good to be true. Will we really be able to sit back, put our feet up and let our vocal cords do the typing?

Raymond Kurzweil of Kurzweil Applied Intelligence Inc. of Massachusetts is an expert on artificial intelligence machines. He is already selling a working system that, in a limited way, operates like a voice-activated typewriter. It's called KVR (Kurzweil VoiceReport) and works like this.

The operator speaks into a microphone, inserting brief pauses – of at least one-tenth of a second – between words. This is called 'discrete speech'. The words are then displayed on a computer screen for verification or editing. If something is wrong, you say 'Scratch that' and the incorrect words are erased. When the dictated text is ready to be put on paper, you simply command 'Print report'.

First, however, each person who will use KVR must 'train' it to recognize his or her voice. That's because KVR is 'speaker-dependent': it will only recognize the speech of users who first pronounce each word of KVR's vocabulary into the system's microphone. The system's computer creates a digital record of the way the user pronounces each word, and stores that record in its memory for comparison purposes during the actual speech recognition process. KVR is even able to distinguish and use homonyms correctly – for instance 'to', 'too', and, 'two'.

At present, however, KVR has a vocabulary of only five thousand words, so its applications are limited. But Kurzweil has developed a number of specialized software packages that make the most of this relatively small vocabulary. For example, VoiceRAD (for radiologists) and VoiceEM (for doctors working in emergency situations) both eliminate the need for human transcribers by making it possible to dictate, edit and print accurate records in minutes. These pieces of software include a feature called 'trigger phrases': a spoken word or phrase will trigger frequently used sentences or paragraphs to be printed in their entirety.

Robert Joseph, the company's marketing director, feels it won't be too long before a generation of easy-to-use, reasonably priced voice-controlled word processors are in homes and offices. These machines will have an unlimited vocabulary and will be able to recognize continuous speech without the user having to pause between words. Then we'll all be able to put up our feet, relax. . . and dictate.

PORTABLE VOICE-ACTIVATED TRANSLATOR

✓ CONVENIENCE

For the Englishman abroad, the language barrier is about to come down with this voice-operated portable computer translator that runs on rechargeable batteries. Voice – that's what it's called – is a hand-held computer with software that can recognize over thirty-five thousand sentences. You simply speak to it in English – for example, 'Waiter, another bottle of your best champagne, please' – and it will electronically speak the words in French – 'Garçon, encore une bouteille de votre meilleur champagne, s'il vous plait' – or in German, Spanish or Italian.

On an LCD screen, you will see what you said in English, to make sure that Voice heard it right. And by pressing a button you can make Voice repeat the phrase in the foreign language and display it, so you can be sure everyone heard it right.

'Voice makes a lap-top computer with a keyboard look like a dinosaur,' says Steve Rondel, president of Advanced Products and Technologies Inc., the Washington-based manufacturers. 'It fits in the palm of your hand and listens to and acts on your spoken command.' Voice weighs 2.8 lb, is about the size of two VHS cassettes stacked together, and runs for three hours on a charge.

Like the Voice-activated typewriter, Voice is speaker-dependent and will only respond to voices that it recognizes. In about an hour, Voice will lead you through an interview in which it memorizes the way you talk. This information is stored on your personal cartridge. Other people can use the machine by going through the same process.

It took six years and millions of dollars to develop Voice, and the researchers aren't finished yet. Its vocabulary will expand, and more languages – specifically Japanese, Chinese and Russian – are on the drawing board.

Rondel sees Voice's first applications at the wealthier end of the tourist industry and in the business community – perhaps this is the way for British business to approach the Single European Market!

WATCH PAGER

✓ CONVENIENCE

Since the Selection telephone isn't with us yet, for the moment we are all still relying on our telephone answering machines. But no answering machine can deliver messages instantly, and a thick, black-box pager around the waist is not everybody's idea of chic. Help, however, is at hand – soon one product will alleviate both problems. It's a digital wristwatch that serves as a sophisticated miniature paging system.

When someone calls, your watch won't embarrass you by beeping loudly at a bad time. Instead, it will gently flash one of several messages – call home, call the office, call the following number – or it will flash a code from 0 to 9 for best friends and other frequent callers. It could take as little as fifteen seconds from the time a message is initiated until it gets to your watch. Since the technology contains both letters and numbers, it should eventually be possible to extend the range and length of messages received.

The watch, to be made by Seiko, contains a receiver designed to function in all countries without modification. The service will be introduced first in the USA, and in other countries very soon afterwards. As production increases the watches are expected to follow the well-worn path of many inventions and become increasingly cheap.

PRAYER WRISTWATCH

✓ CONVENIENCE

Here's a useful little device for another sort of traveller. If you're a Moslem, arranging your necessary prayer schedule while on a business trip abroad can be a problem. David Kohler, an Englishman who worked in Saudi Arabia, recognized the need for a handy computerized system and has come up with a special kind of watch that lets even the fastest-moving businessman keep track of the five daily prayer calls and the holy days while flitting through different time zones.

Powered by two tiny microprocessors and with a 12K memory, this amazing timepiece points to Mecca (which the worshipper must be facing when praying), chimes ten minutes before prayer calls (to enable the worshipper to wash before praying, as stipulated in the Koran), and flashes the date according to the Moslem calendar.

This is not an easy task when one considers all the variables. To set the watch, the wearer must programme in the time, the date and his position on the globe. Two hundred locations are set in the watch's memory, so if you're in one of those places, it's automatic. If not, the watch will help you work out what to input. Since prayer times are calculated differently in different time zones of the world, the watch has five programs from which to choose in determining the correct prayer time for the wearer's location.

PANIC ALARM WRISTWATCH

SECURITY

Considering the crime rate and the fact that watches today can do just about anything, it's surprising that this kind of alarm watch didn't arrive years ago.

The principle is really simple. A battery-powered siren is set off when the wearer opens the wristband or an attacker knocks it loose. The watch comes off easily as the band has a Velcro fastener.

The inventor is Nathan Feigenblatt, a retired safety engineer from Tucson, Arizona. He patented the watch back in 1986 but so far, unfortunately, has not found a manufacturer to produce it.

3
LEISURE 2001

Digital audiotapes and decks
Digital speakers
Compact disc recorder
Private listening at home
3-D sound
Larger-than-life TV
High-definition TV
Flat TV
Smart TV
Talking VCR remote control
Video-on-the-go
Desktop video
Virtual World
Motion simulator
Super movies
Dial 'M' for movies
Binocular glasses
Spy satellite photos
Electronic still photography
Memory card camera
Camera stabilization lens
Oversized golf
Night golf
Perfect line golf ball
Simulated golf
Sweet-spot tennis racquet
Line judge tennis balls
Swimming propulsion device
Water walkers
Uphill skiing
Nail-less horseshoe
Sports shock meter
Punch meter

DIGITAL AUDIOTAPES AND DECKS

ENHANCEMENT

An American scientific magazine has called digital audiotape (DAT) recorders 'the finest home tape recorders ever invented'. The problem is that the record companies fear a massive outbreak of illegal copying and piracy. Consequently they have been lobbying for these new tape decks to be equipped with anti-copy circuitry; otherwise their profits will take a nose-dive as home audio enthusiasts make duplicates of compact discs and pre-recorded tapes.

But – with or without the capacity to erase and record – digital audiotapes and decks are undoubtedly coming. And that's good news for everyone who loves music, for DAT quality is outstanding. Indeed, many audio experts claim that every tape is as good as the master made in the recording studio.

What's more, DATs hold more music. Each tape, only a little larger than a matchbox, can store up to two hours' worth of sound. And you can scan a tape in a mere forty-five seconds, so you don't have to wait what seems like forever for the deck to track down your favourite song.

Part of the secret of DAT's speed and performance is the way in which information is arranged on the tape. Standard tapes record along their length, whereas DATs record along their width. It's more efficient – like stacking books the normal way along a shelf, instead of end to end.

DAT decks are a cross between a VCR and a standard audiotape player. All the advanced functions, such as auto search and programmable playback, are exactly the same as in the old players, but the tape loads like a VCR. Also like a VCR, DAT recorders require a few seconds to wrap the tape around its record/play drum.

Another major advantage of DATs is that they are not nearly as sensitive to bouncing and jarring as compact discs. The slightest bump while driving or jogging will cause a CD to skip. Not so with DATs, which makes them ideal for pocket-sized portables and cars.

DIGITAL SPEAKERS

ENHANCEMENT

Now that digital audiotapes and decks are on their way, you'll need to be more discriminating when selecting loudspeakers. How good your system sounds has always been highly dependent on the quality of its speakers. After all, it's the speakers' responsibility to convert electronic signals from the amplifiers into the sound waves that eventually enter your ears.

Digital recordings capture a wider range of frequencies and a greater dynamic range – the difference between loud and soft passages – than do conventional recordings. But the overall effect is lost on speakers incapable of reproducing the sounds accurately and duplicating musical peaks (especially loud bass passages) without distortion or audible vibration. This is the challenge facing loudspeaker manufacturers.

JBL Inc., a California company, is experimenting with titanium – an ultra-strong, lightweight metal. Its strength-to-weight ratio allows the speakers to react more quickly and accurately to the electronic signals. Hector Martinez of JBL says, 'It provides more clarity and acoustic articulation' – in other words, it sounds terrific.

Two British firms are also hard at work. KEF Electronics is aiming to eliminate the cabinet rattle caused by digital's increased power; the company is connecting the two woofers in each cabinet – the large speakers responsible for bass and lower-frequency sound – with a stiff metal bar to help cancel out vibration. B&W Loudspeakers has gone a stage further and is busy testing a 'cabinetless' speaker whose design incorporates a honeycomb-like housing filled with acoustic foam.

Soon the shops will be stocking these and other innovative speakers. Although they've been called more evolutionary than revolutionary, when it comes to great sound quality their importance can't be denied.

COMPACT DISC RECORDER

✓ ENHANCEMENT

Compact discs may sound better and last longer than audiocassette tapes, but they still have one disadvantage. You can't erase and record on a CD. But that's going to change soon. In the USA, Tandy Corporation has applied for several patents on a compact disc record-and-play technology that is similar to a VCR. You will be able to erase, record, play and even edit your discs.

To put it simply, a CD is pressed just like an old-fashioned record. If you look closely you can see the grooves. Those grooves have pits and bumps that contain information. A laser scans the disc and converts the information to sound. Now Tandy has created a new disc called THOR-CD, comprising several different types of layers, one of which is heat-sensitive. When heated by the laser, the pits and bumps literally melt away.

Another layer is reflective and lets you create new pits and bumps when you want to record. There is also a layer that protects against fingerprints and dust.

This new recorder will be 'completely backward compatible', according to Robert McClure, president of Tandy. This means that the CDs you already own will play on the new machine – but you won't, of course, be able to erase or record over them.

PRIVATE LISTENING AT HOME

●●● ✓ ENHANCEMENT ✓ CONVENIENCE

Just when you thought there couldn't possibly be another small portable listening device, along comes Private Waves, a new type of wireless headphone from Datawave Inc. of California. But don't be cynical – this one does something none of the others can do.

The sounds you hear over Private Waves can originate from your stereo, VCR, CD player or TV. So you can be making beds, washing the car or digging the garden – and still listen to your favourite music or TV shows. Like a Walkman or mini-cassette player, it has a miniature receiver that clips on to your belt or goes into your pocket. Lightweight bud-style headphones are connected to this receiver.

Other so-called wireless headphones exist, but they are restricted to line-of-sight use. That's because they use infra-red technology. Private Waves, on the other hand, relies on radio frequency (RF) transmission. Since RF can travel through walls, which infra-red cannot, you can use these headphones anywhere. Well, anywhere within 75 feet of the source.

The Private Waves system is easy to set up. A compact transmitter is connected to the audio-out or headphone jack of your TV, VCR, stereo or CD player. You just clip on a 3½ oz mini-receiver, put the headphones on your ears, and you're in business.

3-D SOUND

ENHANCEMENT

Two inventors with impeccable credentials – one a former long-time employee of IBM and the other a conultant to NASA at 18 – are about to revolutionize the way in which we hear our entertainment. Ralph Schaefer, president of American Natural Sound Development, explains that current stereo audio is delivered in a lateral wall of sound. But with 3-D sound it will be all around us.

It's different, too, from quadraphonic sound, which requires multiple channels of sound recorded from lots of microphones and then transmitted through speakers. With 3-D sound, only one piece of recording equipment is necessary for production, and the usual two channels of sound for reproduction. Since the difference lies in how the sound is recorded, there's no need to buy any new playback equipment.

Schaefer and his partner Peter Myers will first market the technology for use in simulators. Then their march into the recording industry will begin – followed shortly, they hope, by an invasion of Hollywood.

LARGER-THAN-LIFE TV

ENHANCEMENT

A TV set with the dramatic impact of a cinema screen has been every electronics manufacturer's dream for many years. Now, the dream looks like becoming reality.

The problem up until now has been that big images always suffered the drawbacks of an over-large set or a poor-quality picture. Now Texas Instruments have come up with a silicon chip that can produce the goods: large, high-quality images from a projection box the size of a pocket camera.

The cathode ray tube that has been the basis of all TV sets since television was first invented uses only one beam to produce all the dots on the screen. TI's Deformable Mirror Device (DMD), on the other hand, can handle thousands of light beams simultaneously, producing much greater detail. According to TI, the new device is also extremely reliable and not susceptible to wear and tear.

Other applications to which the new technology will probably be put include target-spotting by military aircraft, laser printing and signature verification.

HIGH-DEFINITION TV

✓ ENHANCEMENT

Television manufacturers have brought us from old black-and-white sets to colour, from models that looked like mahogany sideboards to miniatures worn on the wrist, and from unruly rabbit's-ear antennas to cable. Now, at last, they're going to do something to improve the quality of the picture.

The goal for High-Definition TV (HDTV) is to look like 35 mm film – rich tones, clarity and all. That means no more fuzzy edges and shallow saturation for the tube of tomorrow.

In order to get this next generation of TV viewing, you will have to buy a new set. As a result, the change from old to new will take place like the transition from black-and-white to colour did – slowly. The new sets will be both wider and bigger – HDTV can only be viewed on screens of 25 inches and over. And since conventional broadcasting is totally different from the HDTV signal, the new sets will have a way to switch back to the 'old' signal, at least for a while, like AM/FM radio.

In the USA it will be available at first only from cable and satellite companies. The networks must wait while their governing body deliberates on nineteen viable proposals and decides which will best serve the viewing public. It has to take into account matters such as band width and lines of resolution, and the all-important compatibility issue: in other words, if broadcasters turn to HDTV en masse, will old sets still be able to receive any signals at all?

FLAT TV

✓ ENHANCEMENT ✓ CONVENIENCE ✓ SPACE-SAVING

Forget about high-resolution and vast screens for a moment. Here's news of a development that lots of us, especially if we live in houses or flats with small rooms, have been wanting for a long time – a TV set so slim that it can hang on the wall like a picture.

The technology is there, but a few problems need to be overcome before that space-consuming telly can be slimmed down and hoisted up on to the living room wall. Sony already markets a battery-operated, hand-held, LCD-based colour set with a 2.7 inch screen. The obvious next step is to make larger LCD panels. But the larger the panel, the higher the rejection rate due to imperfections. Large panels are also prohibitively expensive at present for use in the domestic market. Until Sony or some other company finds a solution, we'll just have to keep hoping.

Meanwhile, there's news that Matsushita, a rival Japanese company, has been testing a new flat screen which combines both LCD and cathode ray tube technology to produce far brighter pictures. Perhaps the breakthrough will be in this area.

SMART TV

✓ CONVENIENCE

So you've missed *Miami Vice* again, and you're hopelessly out of touch with *Neighbours*. Perhaps if you could find where the kids have hidden the TV papers you might at least manage to ask them (not for the first time) how to programme the video to come on at the right time . . .

Relax! In future, when you feel like watching the tube, you'll just pick up your SmarTV remote control and press ON. The SmarTV knows just what you like and has been recording your favourites all week – it can do up to 250 shows at a time. Just select which one you want to see from the jukebox-like menu that comes up on the screen. Another button will give you a brief description of the show. If you think you're interested, press PLAY SHOW. You never have to touch a tape, or figure out how to programme the video. You can always stop the show (say when the phone rings), rewind (if you miss a joke) and, best of all, you can fast-forward at twenty-seven times regular speed past the commercials. Or press STOP and check the menu again.

Blair Newman, of Metaview Corporation in San Francisco, dreamed up SmarTV some time ago. 'I came up with the idea,' he says, 'and when several years later no one was doing it, I decided I'd just go ahead and do it.'

The 'genius VCR', as Newman calls his baby, is a black box about the size of a 19 inch TV set. After you've programmed into it your telly-watching tastes, it monitors all channels at all times of day and records every show it thinks you might like to see.

TALKING VCR REMOTE CONTROL

CONVENIENCE

If you can't afford SmarTV, here's another alternative (likely to be cheaper) that will save you having to grovel to the kids for help with programming the VCR. To take the frustration out of all that time setting and button pushing, Sharp Electronic Corporation has come up with a remote control unit that verbally guides you through the programming process.

The Optonica Voice Coach has a fifty-phrase, five hundred-word vocabulary. Even if you suffer from machine phobia, you'll be able to handle the step-by-step spoken instructions. The Voice Coach also allows you to record the programme you're watching while at the same time recording the audio from another source, such as an FM simultaneous transmission (of, say, a concert) on the radio. That way you get the best of both worlds. The remote unit doubles as a timer, alerting you when your favourite TV show is due. 'Alarm time is 8 p.m.,' it will tell you.

To begin with, the Voice Coach will be available only with Optonica VCRs. But with the increasing sophistication of video equipment, talking controls should become widespread.

VIDEO-ON-THE-GO

✓ ENHANCEMENT ✓ CONVENIENCE

Another bastion has been successfully assaulted in the miniaturization revolution. Sony's battery-operated GV-8 Video Walkman is both a TV and a VCR. It's hand-held, about the size of a paperback, weighs 2 lb and has a 3 inch colour LCD screen. The VCR uses tiny 8 mm videocassettes, which can play or record for four hours in the 'extended play mode'.

The Video Walkman can be used virtually anywhere. In the car, on public transport, while preparing a meal in the kitchen, or lying in a hammock in the garden are just a few suggestions from Shinichi Takagi, president of Sony's consumer video products division.

Business applications abound as well. Sales training tapes could be viewed minutes before an appointment, and product demonstration videos could be shown to clients over lunch. Sony also expect that magazine publishers in other countries will take the opportunity, as their Japanese counterparts have already done, to start marketing video versions of their publications.

DESKTOP VIDEO

✓ ENHANCEMENT ✓ CONVENIENCE

Soon after desktop publishing became a buzz phrase in the computer world, desktop video rolled itself on to the list of the new and amazing. The professionals who do the computer graphics on the TV were astonished at first, but for basic video enhancing many of them now have in their studios the same system that you can buy for home use. Now it's your turn to make the clouds speed by under images or words that you create.

The system, manufactured by Commodore, is called the Amiga 500. With one of these hooked up to the VCR you can take tapes of your wedding, your children's christenings, Christmas with the family and so on, and add sound, graphics, titles and other touches for a professional broadcast flavour. If you spend a bit more you can upgrade the Amiga's genlock, the gizmo that synchronizes the videotape signal with the computer-generated graphics. The better genlocks give your home production a sharper image, with higher resolution.

As with all computer systems, a lot depends on your software. But fear not – plenty exists that will enable you to re-create those technical tricks with which we've become familiar: rolling, twisting and turning of images, as well as dissolves, fades, wipes, squeezes, page flips, animation and the full range of computer graphics in various typefaces and colours.

VIRTUAL WORLD

●●○ ✓ WACKY

Here's a home entertainment that gets right away from new developments in audio, stereo, computer graphics and so on and makes them seem quite commonplace. In fact, it's out of this world. What we're talking about is the ability to visit faraway places or even different eras without leaving your living room.

Suppose you want to visit a tropical island. First, you'll put on a headset that replicates the sound of the surf, the wind and the wildlife. Next, you'll don special goggles with visual displays like a mini-TV. Not only will you get a three-dimensional look at the island, but as you turn your head the scenery will change, giving you a panoramic view of your surroundings. Finally, climb into your Virtual World suit and gloves, which let you feel the heat, the breeze, the sand, the textures of anything you might touch on the island. All of this equipment is hooked up to the 'home reality engine', an egg-shaped unit that plugs into an ordinary phone jack.

For the final bit of magic, any friend also wearing a Virtual World outfit can join you from anywhere on earth on this tropical adventure.

You can experience many sensations and visit many environments in the Virtual World. Once you are kitted up, 'previews' will appear in your special goggles. All the possible worlds you can visit are lined up before you like aquariums in a pet shop. There might be a Prehistoric World, a Wild Western World and a Future World. You just place your Virtual World glove in the aquarium of your choice and you are instantly transported to that time and place.

But you're not finished yet. At the Virtual World 'make-up mirror' you get to choose the life form you wish to take when you visit these Virtual Worlds. Go back to prehistoric times as a dinosaur, if you like. A friend visiting the same world will see you as a dinosaur. Fancy going back to the Old West looking like John Wayne? It can be done, pardner.

If all this sounds like a modern version of *Alice in Wonderland*, you may not be too far from the truth. The inventor of Virtual World is Jaron Lanier, a student drop-out who subsequently became a computer scientist. 'It feels', he says, 'like being in a computer-generated dream.'

But the concept is already much more than a dream – it exists. Virtual World is in fact a development of systems such as flight simulators. Its more serious applications include enabling architects and clients to walk through building designs, and allowing science students to do the same with fields of molecules.

For those of us with less high-flown aspirations, Lanier expects Virtual World to be available at amusement parks in the early 1990s, followed during the decade by first a high-priced home version, and then a cheaper mass-market one.

MOTION SIMULATOR
●●○ ✓ WACKY

Here's another 'simulation' entertainment for the times when reality isn't quite good enough. Want to go on a roller-coaster ride, but it's raining? Want to go on the Ferris wheel at the fair, but you're afraid of heights? A Swiss company, Intamin, has thought up a solution – just sit yourself down in one of its Dynamic Motion Simulation Theatres.

Here you'll find yourself in what appears to be a relatively normal-looking seat. But when the movie begins, hold on to your cowboy hat! The seat will rise and fall, and shimmy and shake, in perfect synchronization with the action before you. The roller-coaster on screen will have a whole new authenticity. When it dips at 60 miles an hour, you'll dip with it enough for your stomach to feel as if it's in your mouth!

The system uses hydraulically activated seats that move in conjunction with a super-high-resolution film technique called Showscan (see next entry) to yield a total viewing experience. Kurt Lukas, vice-president of Intamin, stresses that the system works best with this new film projection because 'Showscan's really the only one that's good enough for the Dynamic Motion Simulator'.

These cinemas already exist in France and San Francisco, and others are on the cards for Japan and Korea. Intamin has plans to install more in American shopping centres, entertainment complexes and amusement parks. Soon, perhaps, we might be seeing them in Britain.

SUPER MOVIES

●●○ ✓ ENHANCEMENT

As television races to look as good as film, the movie industry is hard at work developing its own improvements. At the top of the list is Showscan, a process that involves projecting 70 mm film at sixty frames per second. Traditional movies are shot on 35 mm film and rolled past the projector's light source at twenty-four frames per second. The increased image area of the

70 mm film plus the more rapid frame speed create a larger, more detailed, more lifelike picture.

Showscan was first developed by Douglas Trumbull, the special effects wizard behind 2001: A Space Odyssey and Close Encounters of the Third Kind. Trumbull has been trying to convince studios and cinemas in the USA to convert but some, would rather fight than switch. The larger film requires a modified projector, a better sound system and a larger screen. The projector can be changed back to accommodate the slower-speed film in half a day, if needed. The Catch-22, however, is that cinema owners are waiting for films shot in Showscan before they finance any conversion. Meanwhile, the film producers are waiting for the cinema owners to convert their projectors before they start making Showscan films!

But some are willing to help push the boat out. Aaron Russo, producer of The Rose, is acquiring the rights to produce three films using the new process. Columbia Pictures has bought 6.6 per cent of the stock in Showscan Film, and Paramount has expressed interest.

DIAL 'M' FOR MOVIES

CONVENIENCE

Those video rental shops that seem to have sprung up like mushrooms among the launderettes, fast food outlets and newsagents of even the smallest suburban shopping streets could be out of business by the year 2000. On the way is an American system called Advanced Broadcasting, which replaces the video shop with a home shopping service that delivers movies right into your computer. All you do is pick up the phone and order up to thirty films at a time. That night, they will be entered into a computer that doubles as a TV for viewing at your leisure.

Well, probably not completely at your leisure. You will either have to pay a fee for every film you want to keep permanently, or the film will erase itself after one use.

Professor Stephen Benton of the Massachusetts Institute of Technology predicts that, in the future, 'free broadcasting as we know it will cease to exist and everything will become pay TV' on systems similar to Advanced Broadcasting.

The movies available will include golden oldies, those just-seen-at-a-cinema-near-you, and quite possibly brand-new films – even premieres. You could have the opening night celebrations right in your living room. 'Should be great for champagne sales,' says Benton.

BINOCULAR GLASSES

✓ CONVENIENCE

Can you imagine an invention that would be greeted with equal enthusiasm by opera buffs and birdwatchers? An American ornithologist and former director of the Chicago Academy of Sciences has come up with just such an invention.

More than forty years ago Dr William Beecher set out to design the perfect birdwatching binoculars. He wanted his instrument to have all the power, field of vision, brightness and resolution of the best-made binoculars in the world, but with only a fraction of the weight.

The Beecher Mirage 7 x 30 binoculars are the result of that quest. A shock- and waterproof body encloses an optical system of fourteen lenses – including four made from what is known as rare-earth glass – and eight super-thin mirrors. Multicoating allows 99 per cent of the light to pass through each lens and reflect from each mirror. The more expensive rare-earth glass creates greater depth of focus. And the mirrors replace the weightier prisms found in most binoculars.

But the best part, according to Dr Beecher, is that the binoculars weigh only 3 oz and can be worn like spectacles.

In fact, he says, the Mirage 'disappears before your eyes', because the mass of the binoculars is in a blind spot.

While the shock- and waterproof qualities shouldn't be necessary at Covent Garden, these binocular glasses will undoubtedly give you a better view of the stage more conveniently than will conventional opera glasses. Sports lovers, too, will be able to see the details of play with no trouble at all. If, that is, you can get hold of the glasses. Currently they are available only from the Beecher Research Co. in Chicago, and their inventor, who donates his profits to charity, has so far resisted the idea of using dealers, because 'that would at least double the price to buyers'.

SPY SATELLITE PHOTOS

✓ ENVIRONMENT-FRIENDLY   ✓ WACKY

And now for something completely different... Did you know that anyone – you, your next door neighbour, your boss, the company you work for, your rivals, environmental groups – can order spy satellite photographs? If you get tired of the antics of the fictional James Bond on your video, why not order up some real-life spy material?

Christer Larsson and his colleagues at Space Media Network of Stockholm make satellite photos available to the public. Several countries had thought of the idea but got nowhere with it. Larsson and his team have done it privately with the help of a Swedish billionaire philanthropist who agreed to absorb the losses of over £100,000 every year until the spy photo business gets off the ground. Their goal is a safer environment and world peace.

The first big wave-maker occurred when Space Media Network sold pictures of the 1986 Chernobyl disaster, thus forcing the Soviet Union to acknowledge the catastrophe publicly. But along with nuclear disaster reporting, the pictures have been used to spot Chinese missile sites in Saudi Arabia, giant forest fires in China, and several military installations and nuclear test sites around the world. The photos are analyzed immediately by international scientists, but often nothing is released for as long as a year until Space Media Network is sure of its findings.

Most of the customers are members of the press, but

with the satellite camera's ability to photograph something as small as 30 feet across, businesses might well want a peek at their competitors' factory on the other side of the world. At several thousand pounds per photo you'd have to need the information rather than just be indulging your curiosity, but as more satellites go up the cost of doing business in the spy photo market should start to come down.

ELECTRONIC STILL PHOTOGRAPHY

ENHANCEMENT

Photography – down on the ground, that is – is going digital. The new system will change the way you take, view, select, develop, edit and feel about pictures. The technology is here now and so is the first generation of cameras, playback drives and printers.

How does it work? Instead of exposing film to light (the way it's always been done since 1839) a chip, similar to that in your camcorder, receives the information framed in your camera lens and records it on a miniature video floppy disc. For the time being, the disc can only be used with the electronic photography hardware.

The main advantage is instantaneous feedback. Film has to be printed before you can view the results of your creativity – or lack of it – but with these floppies you can hook up a special playback device to your TV set, pop in the disc and instantly preview your pictures before deciding which of them are worth printing.

You can preview the pictures at any time, whether

you've shot one frame or the whole twenty-five or thirty available on the disc. With the help of a pricey machine you can print your own photos at home. Alternatively, send your disc off to the developers and mark the frames you want them to print and return.

Unfortunately, the current state of the art needs more work at the drawing board. The quality of the pictures is poor (they've been likened to a colour Xerox) and the cost of the equipment high. It'll be much later in the decade before the system is perfected and the equipment is in our high street photographic shops.

MEMORY CARD CAMERA

●●○   ✓ MONEY-SAVING   ✓ ENHANCEMENT

Forget about film. Forget about video floppy discs. We may be bypassing these still picture systems altogether in favour of something even newer and better. Second-generation electronic cameras dispense with the floppy disc in favour of a credit card-sized 'memory card' which has no moving parts, stores digital images in a solid state memory, and creates a much higher-quality picture.

In brief, the IC Memory Card camera works like this: a charged couple device (CCD) converts light photons to electrons (electrical energy), and the pictures are then recorded on the memory card. At present the card can hold twelve 'fields' (pictures), but in the future it should be able to hold forty.

A key feature of the card is that it carries its own tiny battery, which can be recharged and used over and over again. So there's no need to buy new cards (as you would films or discs). After taking your pictures you can preview and edit them on a special playback device. Next, you transfer the images to your personal computer disc or on to digital tape. From the tape, you can make prints using a special printer. When you've finished with that batch of pictures, just go out shooting again. The advantage over electronic still photography (see previous entry) is that there is no loss of quality as you go from tape to player to printer.

Such a system should be of interest to everyone who uses a camera. At present, however, the IC Memory Card camera is far more expensive than its possibly short-lived predecessor and will therefore only be viable for professionals at first. But the manufacturers feel that the general public will be sufficiently attracted by its better pictures to wait a few years for prices to fall, rather than plump for the cheaper, poorer floppy disc cameras that are available now.

CAMERA STABILIZATION LENS

●●● ✓ CONVENIENCE

New lenses that take the shake out of pictures should help professional photographers and amateurs alike. Canon has already shown a prototype of the system in its L series 300 mm lens. Demonstrated on a vibrating platform, it produced rock-steady pictures. The lens is about 10 per cent heavier and about 25 per cent longer than normal.

And the principle involved? Inside the new lens,

highly sensitive meters detect camera vibration and respond with an optical compensator that actually displaces the image so that it appears shake-free.

Although Canon anticipates that the system will be used first in long telephoto lenses (the longer and heavier the lens, the harder it is to hold it steady, and the obvious solution, the tripod, isn't always a practical one), it should eventually find its way to even the simplest snapshot camera. Image-stabilization lenses will also make it easier to get sharp pictures at low speeds, when even the slightest movement can blur a photo.

OVERSIZED GOLF

☑ MONEY-SAVING ☑ TIME-SAVING ☑ WACKY

There's more to entertainment than tweeters and woofers, digital this and liquid crystal that. Now to the great outdoors! Except that sometimes it's hard to get there. Golf is now a very popular game, but many prospective players shy away because of long delays on public golf courses and high fees on private ones. And even if they do get a game, novices are often intimidated by the bunkers, dog legs and other course obstacles that can get in the way of a good time.

So here's a version that's easier and faster to play, on courses that are tiny by normal standards. An American company, Golfun Equities, has come up with a pressurized plastic ball the size of a grapefruit. The game requires over-sized tees, but players use standard clubs. The air-filled ball (available in hot pink and other colours) travels one-third as far as regular golf balls, and, better still, only 6½ acres is required for a nine-hole golf course, compared to the usual 180 acres. It takes no more than an hour to complete a game of Golfun.

'It is intended both as family recreation and as a training ground for regular golfers,' says Jim Contini, joint owner of the firm. The company plans to establish Golfun franchises around the world through contacts with property developers, amusement parks, local authorities and anyone else with enough land to spare.

NIGHT GOLF

✓ WACKY

Here's one for the real golf addicts! A group of entrepreneurs in New Hampshire has dreamed up the idea of illuminated equipment, so you can now play your favourite sport at night.

The balls are translucent and made of polyurethane, with a hole for a 1 inch chemi-luminescent stick that is inserted into the ball by the golfer before playing. The stick emits a green glow that lasts for about six hours.

Pick Point, the company responsible for Nitelite, claims that these strange balls drive, chip and putt just like the daylight white ones. New Hampshire has already played host to nocturnal tournaments with fairways, flag pins and cups all lit by chemi-luminescent sticks. Players use flashlights and lighted trolleys to get around.

Nitelite is great news for golfers; bad news for golf widows.

PERFECT LINE GOLF BALL

✓ ENHANCEMENT

To the novice, every golf ball may look like every other golf ball. They're white and have dimples. But a major American golf ball manufacturer has developed a super ball for precision putting.

To explain how it works, we first need to explain the technique of putting. The experts line up their putts along the ball's parting line, an axis that divides the ball in two. When the putter hits the ball along its 'equator' the ball should run true.

Most designs are based on dimples arranged in equilateral triangles, giving the ball bilateral symmetry (one half is the mirror image of the other). William Gobush, manager of aerodynamic research and testing at Acushnet Co., has patented a design in which the ball's dimples are arranged in twenty-four isosceles triangles (two sides are of equal length). This, in effect, gives the ball six planes of bilateral symmetry. So golfers have a much better chance of stroking true putts. 'When driving,' says Gobush, 'a pro will usually hit the ball's "pole" region as opposed to its parting line. But in putting, the force on the putter is more controlled and he'll get a symmetrical hit if he can find that parting line axis.'

You don't have to be a Faldo, a Norman or a Ballesteros to make use of this new ball, but – let's be honest – if your handicap's way up in the clouds you're probably going to be wasting your money. A reasonable degree of expertise is required before such refinements are worth experimenting with.

TOWARDS 2001

SIMULATED GOLF

●●● ✓ CONVENIENCE

And now, for those who can't get to a golf course, or who merely want to practise all hours to improve their game, here's an invention from the windswept prairies of Canada that is so real, so scientific and so entertaining that it could revolutionize the sport. Its first markets will be Japan, where space for real golf courses is at a premium, and Canada, where cold weather makes the golfing season very short.

The game – GS 2020 Interactive Golf Simulator, to give it its official title – has four components: an interactive video with a large screen; a tethered ball; a scale that measures the weight or power stroke that goes into the golfer's swing; and a set of optical sensors that measure the swing of the club. An actual course is played: the famous Banff Springs Golf Club course in Alberta. New courses will be available at the rate of two a year.

The golfer begins with a view from the first tee seen on a 19 inch monitor. He chooses his club and feeds that information into the computer. He swings, and on the screen actually sees the flight of the ball and the spot where it lands. He picks his next club after he sees the view for the second shot. In determining how far, how high and how straight the ball will travel the computer takes into account the club used, how much power went into the swing, how cleanly the ball was hit and the wind factor.

According to Joytec, the Saskatchewan manufacturers, 'The computer operates a 12 inch laser disc with 54,000 frames.' That means there is enough video for hundreds of possible shots. You can end up in the trees, water or bunkers, and even see wildlife scurrying across the fairway.

Hotels, golf centres and clubs are the first places likely to offer the GS 2020. But the entire thing needs only a 12 x 12 foot playing area with a 10 foot ceiling, so Joytec expects to be selling a version of the Golf Simulator for home use.

SWEET-SPOT TENNIS RACQUET

●●● ✓ ENHANCEMENT

If you've ever played a game of tennis, you probably know the 'sweet' sensation of hitting the ball with the centre of the racquet. The centre, responsible for the sound and feel of that characteristic 'whomp', is understandably referred to as the 'sweet spot'. You've probably also felt the momentary twinge that shoots up your arm when you don't hit the ball so well. The strain to your tendons causes painful tennis elbow.

Now a company called Dynaspot is making a racquet that puts a sweet spot in every shot and reduces the forces that cause tennis elbow by about 40 per cent. Marvin Sassler, president of the company, is an electrical engineer and tennis player who once had a bad case of tennis elbow. While recuperating, he came up with the idea of a racquet with a sweet spot that moves.

Gerald Geraldi, Sassler's chief engineer, designed a racquet frame of lightweight graphite with partially filled tubes of liquid on either face that absorb the torque when the ball is hit. The fluid acts as a 'dynamic' balancing system, meaning that the weights move, thereby cushioning the impact of the ball and reducing the stress on the player's muscles. The energy in your swing is delivered to the ball, and not absorbed by your arm.

Sassler's revolutionary racquet has been spotted at both the French and Geneva Open Championships. Will Wimbledon ever be the same again?

LINE JUDGE TENNIS BALLS

ENHANCEMENT

Cheaters beware! In the future, you won't be able to win that desperately needed point by calling a close ball out.

Engineer John A. Van Auken has developed special tennis balls, called Accu-Call balls, that contain electricity-conducting metallic fibres. On standard tennis courts, these balls perform the same as any others. But when used in conjunction with special sensor-embedded courts, they can make line calls themselves. Every time the ball makes contact with the court, a signal is emitted to a laptop console held by a line judge at the side of the court. This signal indicates to within one hundredth of an inch whether the ball landed inside, on or outside the line.

So far the balls have been successfully tested by John McEnroe and Jimmy Connors. The only (minor) drawback is that the metallic fibres cause the balls to be slightly grey in colour. But Penn Athletics, manufacturer of the balls, says this problem can be corrected.

At first the balls will only be available for hire at private clubs equipped with the special courts. At some point in the future, however, the public will probably be able to buy them from sports shops for use on private or public electronic courts.

SWIMMING PROPULSION DEVICE

☑ WACKY   ☑ CONVENIENCE

They look like two torpedoes. You strap them on, stretch your arms, flick a switch, and the next thing you know you're jetting through the water like James Bond testing out Q's latest box of tricks.

The (at present nameless) propulsion device can do 4 miles an hour for forty-five minutes in a lake or pond or in the sea. Each of the units includes a battery, a motor and a propeller in a watertight housing. Within easy reach of your fingers is a control panel with an on/off switch. One model will also have a light for night swimming. The contoured units strap snugly under the arms and extend from the hand to the elbow.

The device was invented and patented by California diver Michael Borges, who is still looking for a company to produce and market it.

WATER WALKERS

☑ WACKY

'To walk on water' has long been synonymous with the impossible. For fifteen years an American named Ben Watson has been trying to do exactly that, and now he seems to have done it. He's obtained a patent and is now looking for a manufacturer. In the meantime, he's taking his water walkers for test runs on a local lake at about 5 miles per hour.

His creation looks a bit like a pair of bulky skis (6 feet long and 9 inches wide), except that each walker is 9 inches deep and the user's foot is inserted inside a foot well instead of a boot. Each walker (or float, to be more accurate) is equipped with eight flaps on the underside to assist forward motion. The walkers are made of foam and fibreglass, and a pair weighs 30 lb; they can support a load of up to 300 lb. The actual motion used is more like that of skating or cross-country skiing than walking, as the floats never leave the water.

The obvious question is: who wants or needs to walk on water enough to spend money on a special device like this? Watson claims his water walkers are stable enough for fishing or hunting, and provide a challenging aerobic off-season work-out for skiers and other athletes. And there would be a certain head-turning excitement about walking across a lake on a bright summer day. And for added fun and excitement Watson has also devised sails to convert the walkers into a catamaran.

UPHILL SKIING

✓ CONVENIENCE

If you're a skier, how many times have you skied down a challenging slope only to be faced with the bigger challenge of getting back up to the top for another run? John Stanford and Phil Huff decided to use their parachuting and skiing experience to design a product that would end that misery. The result is a lightweight parachute powerful enough to propel skiers up steep slopes, yet small enough to be easily packed away for the trip back down.

Coming up with a prototype was fairly simple – since Stanford's company manufactures parachutes – but testing it was downright thrilling. 'We realized that uphill skiing was more fun than skiing downhill,' says Stanford. So once they got a patent registered, the sport of 'upskiing' was born.

The parachute can be used on snow-covered lakes or steep mountain slopes, in winds as low as 7 or 8 miles per hour. Stanford and Hugg have upskied in 50 mph winds, but describe the experience as 'terrifying' and 'dangerous' and strongly advise against it.

Like sailing, upskiing is a wind sport. After putting on your skis and strapping yourself into the harness, you lift part of the canopy (with the help of control lines) so that it fills with wind. Then you just lean back and go. On a modest slope with firm snow and average winds of about 10 to 15 mph you can move at twice the speed of the wind. To get yourself up a very steep slope, however, you'll need stronger winds of about 18 to 20 mph. A control centre attached to the harness allows you to increase or decrease your speed and, in the event of an emergency, to release yourself from the equipment. The parachute itself is 28 feet in diameter; the whole system weighs a mere 13 lb and folds up to the size of a backpack.

Uphill skiing is not, of course, restricted to intrepid Americans. It has also been demonstrated in the Cumbrian Fells and on the Atlantic coast of France.

NAIL-LESS HORSESHOE

✓ CONVENIENCE ✓ HEALTHIER

And now here's something for those who like to spend their spare time on horseback – or, at least, for their mounts. At long last, someone has come up with a new line of footwear for the stylish horse.

The fashion – a bold departure from the traditional horseshoe – is made of plastic, and is only a third the weight of steel. To provide greater comfort, nails are dispensed with in favour of an adhesive. The plastic shoe – durable, high-impact and flexible – is heated and moulded to the hoof for a perfect fit.

The shoe was designed by two American dentists after a horse-breeding patient had complained, in the course of conversation, about the detrimental effects of traditional steel shoes. They were unyielding, he told them, and didn't allow the hooves to expand as they should. On top of that the nails dried out the hoof, cracking and weakening it.

His two listeners decided to apply the principles of dental bonding to the problem, reasoning that a hoof isn't all that different from a tooth – both need the same protective care to last and to perform well. Their invention, they say, ensures the integrity of the hoof, and even strengthens the hoof wall and protects it from premature wear. As a result, far less strain is placed on the horse's tendons and ligaments.

The new shoe, which is already on sale in the USA, is not intended to replace steel shoes entirely, but to be used every third time to enable the nail holes to grow out for the next steel shoeing. As with many inventions, the price should become very competitive as the idea catches on and more sets are sold.

SPORTS SHOCK METER

●●○   ✓ SAFETY

And now to the down-side of sport – the injuries and shock that it may inflict on your body. John Carlin, a Denver entrepreneur and sports fan, has patented a device that he believes could make all contact sports safer. Called the Measureguard, it gauges the accumulated shock sustained during the course of a game.

In effect, this meter will tell managers and physiotherapists how 'used' a player has become and will reveal his susceptibility to certain kinds of injuries. The Measureguard, attached to the back of American football helmets or other clothing, monitors the shock with a vibration sensor. Each jolt produces electrical signals that are converted to numbers and displayed on the tiny screen attached to the helmet. The signals fade as time passes, but increase with each new blow.

'An official can see who is over-used – and thus prone to injury – just by looking at the back of the helmets,' says Carlin. In his view, the Measureguard could prevent many tragic injuries among both schoolboys and adult players (in Britain, rugby is one of the main culprits). In fact, he came up with the idea after his sixteen-year-old nephew was hurt while playing American football.

PUNCH METER

●●○ ✓ CONVENIENCE ✓ SAFETY

Apart from the Measureguard (see previous entry) John Carlin has also patented the Measurehand, a device that measures the effectiveness of punches thrown by a boxer. Vibration sensors worn around each wrist gauge the shock when fists meet target. This information is transmitted by radio signal to a processor that records the number of blows, the time of each blow, the frequency, force and accumulation of force. The information is displayed digitally on a TV screen. Using this feedback, trainers and boxers can determine which combinations, punches and techniques produce the most devastating results. Also, spectators could appreciate boxing more by knowing instantly the effectiveness of each punch.

To determine both give and take, the boxer can also wear the Measureguard to gauge the amount of shock absorbed by his body during the bout.

4
HEALTH MATTERS IN THE FUTURE

Cholesterol-reducing powder
Gyro exercise machine
Aquatic exercise machine
Holograph bifocal contact lenses
Time-release eye drops
Eye braces
Laser treatment for sight defects
TV eyeglasses
Super fish-eye lens
Smart contact lens
Digital hearing aid
Digital implant for the totally deaf
Talking glove for the deaf
Anti-cavity pill
Painless dentist's drill
Computer-designed dental crowns
Mammogram without X-rays
Stress gum
Non-invasive home glucose test for diabetics
Fever check
Flexible bone clamp
Pregaphone
Sonic painkiller
Microwave painkiller
Smart pill bottle
Electronic Bandage
Electronic deodorant
Sunburn protection meter
Kiss moisturizer
Living Skin Equivalent
Super pore cleaner
Wrinkle-reducing pillow

Snore-reducing pillow
Electrified birth control
Time-release birth control
Impotence pills
Radon extractor
Phone shields
Vilest taste
Diagnostic loo
Freezing humans

CHOLESTEROL-REDUCING POWDER

✓ HEALTHIER

Common sense and the escalating cost of providing health care dictate that nowadays we pay a lot of attention to preventative health – to diet and exercise. We all know about the connection between cholesterol and heart disease, but scientists at an American university have nearly got that one licked. 'Ten years from now we'll have shakers on the table for salt and pepper – and one for the cholesterol slasher,' says Donald Beitz, head of the research team at Iowa State University which is developing a method of removing cholesterol from animal products.

They've already established that eubacteria, a natural substance found in the body, contains an enzyme that alters cholesterol so that it is absorbed poorly. Once the cholesterol is converted into this state, it would pass right through the body without stopping at the hips for a few months – or in the arteries for ever. The enzyme could eliminate as much as 80 per cent of the cholesterol that we absorb from our food.

The trick is to isolate the enzyme and reproduce it in quantity, probably through genetic engineering. It could be sold as a condiment for use at your table, or it could be used in pre-packed foods such as tinned meats that are available on supermarket shelves. Remember sausages, bacon and eggs? The traditional British breakfast may yet make a comeback!

GYRO EXERCISE MACHINE

●●● ✓ ENHANCEMENT

The other half of the 'healthy living' equation is exercise – but the same routine day after day can become very boring. So forget those tired old exercise bikes and rowing machines. How's this for a bit of stimulus to get you bending and stretching again?

Picture three giant concentric hula hoops standing on end. Now imagine yourself strapped into the innermost hoop. Move a muscle and three connected rings start to sway. Strain a little harder and you begin to spin. Nod your head and you somersault right over. You may already have recognized the principle – the design is based on the gyroscope, a compass device that's been around since the 1850s.

Gyro stands 9 feet high and wide. The three rings – one green, one red, one yellow – are made of tubular steel and rotate around one another, each on its own axis. The rider is fastened into the innermost ring with a footbinding system and a padded waist device that allows for minimal movement. There are handlebars overhead for easy access and control; they also help stretch and support the body.

Once you are in place, the slightest body movement will get the whole system turning. Through subtle shifts in weight and isometric muscle contraction, the rider can create and control the action, speed and duration of the exercise. The rings move in every direction, the outer ones keeping the whole apparatus in balance. It's possible to do forward dives, back flips, lateral rotations, cartwheels and more, all with the weightlessness of an astronaut in space.

Gyro North America says, 'It's an exciting and exhilarating workout.' It's also safe, and puts minimal stress on the joints. In spite of what you might think, this is no fairground roller-coaster and no one gets motion-sick. You can slow the machine down or stop it and return to an upright position simply by bending your knees, which lowers your centre of gravity.

AQUATIC EXERCISE MACHINE

●●● ✓ ENHANCEMENT

If you feel your natural exercising element is water, this is for you. It's a swim-in-place mini-pool that will do for swimmers what the treadmill does for runners – and with less possibility of injury, because exercising in water (as opposed to air) reduces the stresses on your body.

The brainchild of an American company, SwimEx

Systems Inc., the machine is just over 17 feet long, nearly 8 feet wide and 5 feet deep. The rear section has a paddlewheel – separated from the swimming area by a heavy grating – that propels water forward in a channel under the bottom of the swim area. The constant current of water then resurfaces when it enters the front of the swim area and flows back against the swimmer. The current can be adjusted to twenty-five different speed settings to meet the individual requirements of any swimmer, but no matter how hard you swim, you won't get anywhere – except fitter, that is.

Since it costs about the same as a decent family car this is not for the average household, but the manufacturers are optimistic that they will sell successfully to sports clubs, gymnasiums and commercial institutions.

HOLOGRAPH BIFOCAL CONTACT LENSES

●●● ✓ ENHANCEMENT ✓ CONVENIENT

However well we look after our health, few of us reach middle age without needing to wear glasses at least some of the time. If you need different glasses for different activities you probably feel that contact lenses, though more flattering to most people, aren't for you. The bifocal contact lenses currently on the market work by dividing their surface into two separate lenses. They don't allow for a full range of vision, and they create a blurry image under certain conditions.

But times are changing. In Britain, Pilkington has been working on a different kind of bifocal contact lens that uses holographic technology. We're all familiar with holographs from the double images on our credit cards. These new contacts, called Diffrax, are etched with concentric rings that split incoming light and also make two images that have equal intensity but different focal points. Your brain will instinctively pick the right image on which to focus, depending on where you're looking.

The Diffrax lenses have been tested successfully for two years. They are easily fitted and take about a week to get used to.

TIME-RELEASE EYE DROPS

✓ CONVENIENCE

Anyone who's had the misfortune of eye problems knows that administering the cure – usually in the form of eye drops – can be pure torture. Half the time you miss, and when you finally do get a drop in, it stings so much that your eyes well up with tears. When they overflow, up to 80 per cent of the solution runs down your face or, worse, the back of your throat. Besides being unpleasant for the patient, this catch-as-catch-can administration is a serious problem when it comes to determining the right dosage. But here's the soluble ophthalmic drug insert, SODI for short, galloping to the rescue.

The inserts are made of a biodegradable polymer matrix permeated with various drugs used to treat an assortment of disorders, from simple infections to glaucoma. SODIs are roughly the size and shape of a Tic Tac mint, and (contrary to what you might think from that information) are easier to insert than contact lenses. Once they're in position under the lower eyelid, they're virtually impossible to feel. The plastic is designed to dissolve over a period of twenty-four hours, steadily and predictably releasing medication. So patients don't have to take drops (effectively or not) ten or twelve times a day; and, because the drug is reliably delivered, drugs can be used in much smaller concentrations and are thus less likely to cause irritations of their own.

The implants were invented by Russian scientists in the late 1970s to treat a disorder common among cosmonauts – conjunctivitis, caused by intense glare in space. In zero gravity drops are useless, but inserts were found to work beautifully.

EYE BRACES

●●○ ✓ CONVENIENCE

If this idea catches on, soon glasses and contact lenses could be things of the past.

Intra-corneal rings work much the same way on eyes as braces do on teeth. Put simply, a thin corneal ring will flatten the cornea to correct short-sightedness. A tighter ring will steepen the cornea to correct long-sightedness. And any ring at all will round out the shape of the eye to correct astigmatism, caused by an oval eye shape.

According to KeraVision Inc., the American

developers of this innovation, the rings will be surgically implanted on an outpatient basis. They can be removed at any time, will cause no interference with normal eye functioning, and could stay in place indefinitely. If the rings (currently under test) are approved for human use, you can say goodbye for ever to mislaid glasses and lens discomfort.

LASER TREATMENT FOR SIGHT DEFECTS

●●● ✓ CONVENIENCE

Here's a surgical and permanent approach to the problem of less than perfect sight. It 'sculpts' the cornea to a better shape. Lasers are used – the kind that cut not by generating heat in the conventional way, but by a photo-chemical reaction. The laser is used for as little as thirty seconds in an operation that can be performed under local anaesthetic.

This is not the same kind of laser treatment that has been pioneered in the Soviet Union – and which has dangers and disadvantages. In that treatment, known as radial keratotomy, a number of cuts are made in the cornea to flatten it and thus reduce short-sightedness. The new method, on which research is being carried out at St Thomas's Hospital in London, is called photo-ablative keratectomy; here, use of lasers separates the chemical bonds holding the tissue together.

However, there may still be hurdles to overcome with the new technology. 'It is exciting, but we have to establish whether it is safe, whether it causes scarring, and whether the induced change in the cornea is stable,' says consultant ophthalmologist Kerr Muir. For this reason the researchers are still cautiously talking about their work as experimental; but companies which manufacture the equipment in a number of countries, and people with sight defects, are eagerly awaiting the results.

TV EYEGLASSES

●●○   ✓ ENHANCEMENT   ✓ HEALTHIER

Sadly, some eye problems are uncorrectable or inoperable. Even here, however, there's hope, for the Wilmer Eye Institute at Baltimore's Johns Hopkins University is collaborating with NASA on an ingenious technology that will help sufferers from tunnel vision, blind spots, peripheral field blindness and night blindness. We're talking about televisions so tiny that they fit into spectacles!

Fibre optics, set in the bridge of the glasses, convey the image in front of the wearer to TV cameras and pump it through a power pack worn at the hip or the shoulder. The images are computerized through the power pack and sent back to the glasses, to be displayed on little solid state TV set screens into the lenses. 'There will be one TV for each lens, to allow for binocular, as opposed to monocular, viewing,' says Marie Leinhaas of the Wilmer Institute. 'Most viewing aids today have the patient use whichever eye is better.'

Because eyes move but glasses don't, the researchers are developing infra-red eye trackers to stabilize the images. The images will be tailored to meet the needs of the individual viewer by programming the computerized power pack for light/dark contrast and magnification. 'Instead of looking at the world the way you and I do, the patient will see reality readjusted on the screen to compensate for his or her vision deficiencies,' says Ms Leinhaas.

Researchers still have a few problems to overcome, not least of which is making the glasses lightweight and acceptable to look at. But by mid-decade or sooner the glasses should be available at a price which many pensioners – it's mostly the elderly who suffer the problems which these glasses would alleviate – will be able to afford.

SUPER FISH-EYE LENS

DISABLED

The copilia, a very rare crustacean found in the Bay of Naples and the Caribbean, is the inspiration for a new fish-eye lens that could give millions of people currently registered blind the chance to see. Its usefulness is dependent on the cause of the blindness – as long as the optic nerve and even just part of the retina are functional, the lens works. What it will do is gather light like no other lens ever devised by man.

According to its developer, Professor Jerome J. Wolken, it is designed to take advantage of the way some marine creatures see while living at depths of between 500 and 1500 feet in total darkness. Over millions of years the lenses in their eyes have evolved to gather short flashes of light emitted by their neighbours as they move. 'Using this type of lens held in an eyeglass frame', says Professor Wolken, 'a visually impaired person can begin to see the world.'

The lens, ten times more sensitive to light than a normal camera lens, is actually two lenses in one. The primary lens provides magnification, while the secondary 'fish-eye' lens built behind it, as close to the eye as possible, collects the light. Manufactured in both glass and plastic, the lens is currently being tested with encouraging results at low-vision clinics and schools for the visually impaired in Pittsburgh and Philadelphia.

SMART CONTACT LENS

●●● ✓ CONVENIENCE

Thanks to a breakthrough by Japanese scientists at Ibaraki University near Tokyo we can look forward to blank contact lenses available at chemist's shops that will automatically correct our eyesight on contact. They developed a gel that behaves like a muscle, when an electric charge is put through this substance it contracts. One application is self-focusing contact lenses, but it could also be used in implants inside the body to release drugs on command. The new discovery would be of particular use to diabetics, who normally have to inject themselves with insulin daily.

Beyond these applications in the field of health care, artificial muscle – for this is what the Japanese have created – offers widespread possibilities in other fields of human endeavour. Chemical energy could be converted into mechanical energy. Artificial muscle would be much more efficient than mechanical motors: muscles use over 50 per cent of the energy they consume; traditional engines such as steam turbines and the internal combustion engine only 35 per cent or so. This alternative source of energy could be of great use in space, where conventional power sources are few and far between.

A further development on which Japanese researchers are currently working is the search for what are known as 'intelligent materials'; these are substances that can alter their shape in response to external stimuli. The ultimate aim is to produce materials such as artificial bone that are capable of growth.

DIGITAL HEARING AID
☑ DISABLED

If you've never had a hearing problem, you won't know what an advance this is. People who say they can't hear usually have difficulty deciphering sounds within certain frequency ranges. Conventional hearing aids use tiny microphones that, when turned up, cause every background noise to sound louder as well.

Enter 3M – manufacturers of adhesive tapes – with something called a digital-analog hybrid hearing aid that allows you to select which sounds will be amplified. The

MemoryMate hearing aid is worn behind one or both ears and can be specially programmed for use in up to eight different 'listening situations'. These settings are programmed by a computer that is controlled by your doctor (or whoever sells you your hearing aid) in relation to your lifestyle. By pressing a button on your hearing aid, you can switch from one 'listening situation' to another, to suit your needs.

Bill Schnier, marketing manager for 3M Hearing Health, suggests a listener might want to eliminate street noise, music at parties or background chatter in a crowded room. 'If you're at a concert,' he says, 'you could eliminate some nearby distractions and concentrate on just hearing the music.' Should your hearing requirements change, you can quickly have your MemoryMate reprogrammed.

The next step to help the deaf and hard of hearing will be a device similar to the MemoryMate, but one that adjusts automatically.

DIGITAL IMPLANT FOR THE TOTALLY DEAF

●●○   ✓ DISABLED

Laura is the name of a welcome new device that will enable people with total deafness of the inner ear to hear. Being developed at the University of Antwerp, the technology electronically stimulates the groups of nerve cells that send messages to the brain from the inner ear.

To achieve this, sixteen tiny silicon-platinum

electrodes are surgically implanted into the inner ear. A small programmable speech processor converts sound from outside the air into digital form. This is then passed down a wire to a radio transmitter in an ear plug, and then by radio link to a circular chip implanted under the skin behind the ear. The chip decodes the radio signal and sends it, via a cable through the mastoid bone, to the electrodes in the inner ear.

Perfect hearing is still a long way off at this stage. What the patient will hear is not natural sound, but something like a computer-generated artificial voice. However, by combining Laura with speech therapy treatment the Antwerp researchers believe that hitherto profoundly deaf people will be able to acquire an average vocabulary within a year.

At present the device's chip has to be packed in expensive titanium to protect it from damage by body fluids. If a cheaper alternative is found, Laura should be a viable proposition for many people who have previously lived in a world of silence.

TALKING GLOVE FOR THE DEAF

DISABLED

Trying to communicate with deaf people who have never learnt to speak can make your heart sink if you don't know their sign language. Not only do you regret not being able to understand them, but you usually feel that you are insulting them in some unforgivable way.

So many thanks to Jim Kramer, a postgraduate in the electrical engineering department at Stanford University in California, who's come up with an invaluable contribution to the quality of life enjoyed both by non-speaking deaf people and by their friends and colleagues. Kramer has applied for a patent on what he calls a Talking Glove.

There are sensors in the flesh-coloured 'glove' that read the positions of the fingers as they spell out words using the American Manual Alphabet, also known as fingerspelling. The glove is wired to a Walkman-size processor unit, worn on the belt, that recognizes the fingerspelling of each letter, combines the letters into complete words, and translates them into synthesized speech. A coin-size speaker pendant worn around the neck projects the speech. 'The speaker can be worn unobtrusively under the shirt,' says Kramer, 'and the sound will emerge from in front of you, as if you were speaking.'

The device also comes with a wristwatch that has an LCD screen which displays what the glove is spelling. As soon as a letter is formed by the gloved hand, it will show up on the wristwatch. When a full word is spelt, the word is immediately spoken.

There is also a pocket-size keypad on which hearing people can type their responses. Whatever is typed on to the miniature keyboard is displayed on the wristwatch for the deaf person to read. The watch can be replaced by a mechanical Braille unit for a person who is both deaf and blind.

'I would like to see the Talking Glove become as accepted as glasses or hearing aids,' says Jim Kramer. 'Put it on your night table at bedtime, then put it on in the morning and you're ready for conversation.'

ANTI-CAVITY PILL

☑ ENHANCEMENT ☑ HEALTHIER ☑ CONVENIENCE

We've had special toothpastes, we've had fluoride in the drinking water and painted on our children's teeth, but here's a simple pill that, medical researchers say, will wipe out tooth decay with the same thoroughness that vaccination eliminated polio. It's still some way off, according to Dr Richard Gregory, the American oral biologist undertaking research into the wonder pill – so don't hold your breath. But it looks as though our grandchildren, at least, may never know what a filling is!

The basic premise, explains Dr Gregory, is that the streptococcus mutans – the bacteria that causes tooth decay – has a variety of what are called antigens that collect on the outer wall of the bacteria cells. Now antigens make our systems produce antibodies, which in this case would fight the harmful bacteria. So Dr Gregory and his team are isolating which antigen or sets of antigens would be most effective in the vaccine, to make the body produce the best sort of antibodies.

Successful tests have been carried out on rats and monkeys, which develop cavities at a faster pace than the eighteen months it takes for humans. Further tests have been undertaken on humans. But yet more studies will be required, to ensure that there are no side-effects, before the pill is finally available to the public.

For those who don't like taking pills, the same research will make plaque-fighting milk possible. Cows immunized with the vaccine produce a milk rich in antibodies that work against the bacteria. When the milk was fed to rats, they were found to be protected against cavities. Dr Gregory sees the milk being freeze-dried and used in food products.

PAINLESS DENTIST'S DRILL

●●○ ✓ ENHANCEMENT

For those of us who still have to fear a trip to the dentist, here's an invention to be welcomed. Since there are no 'good' vibrations when it comes to the dentist's drill, no vibrations at all will be a major improvement. Just imagine – no high-power buzzing, no pain, no squirming. In fact, no drilling!

Pfizer Laser Systems is working on the last details of a laser instrument to replace the drill. The company has already been successful with a laser tool for gum surgery. The trick to using the same technology on hard tooth enamel is getting the right amount of power – too much power can crack the tooth.

Pfizer says the laser drill will 'remove decay, seal fissures in an effort to prevent decay, and even seal any open ends after a root canal to prevent infection and/or the need for any repeat root canal'. Great – but will it hurt? 'Patients will feel maybe a slight heat sensation, but nothing else. Psychologically, the patient will love this.'

Stringent US government regulations mean that it will

be some years before the new non-drilling drill is approved for use on the American public. Meanwhile, a Paris dentist is already using a similar instrument. Here's hoping it may one day be available on the National Health!

COMPUTER-DESIGNED DENTAL CROWNS

☑ TIME SAVING

Any invention that saves a trip to the dentist is okay in most people's book and here's one that's actually here now. Researchers at the University of Minnesota have come up with a computer-designed dental crown.

With their system, photographs – rather than impressions - will be taken of the damaged tooth. These will be digitized to create a 3-D model of the crown on a computer screen. Software will then instruct a milling machine to make the crown.

The process is less expensive than the traditional one, because stainless steel and pre-cast ceramics are used instead of gold and unformed ceramics. And, perhaps best of all, the whole process takes about an hour, cutting out the need for a temporary crown and a second visit to the dentist. But unless you have a private dentist who can afford the whole system (well over £100,000) so that he or she can manufacture the crown in the surgery, the details will have to be sent out to a lab and you'll have to come back to have your crown fitted.

MAMMOGRAM WITHOUT X-RAYS

✓ ENHANCEMENT ✓ SAFETY

Checks and tests on our bodies, whether we're healthy or ill, are always good news because they imply that any potential problem can be caught before it develops too far. One woman in ten develops breast cancer at some time in her life, and up until now the best way to detect it at an early stage was to have an X-ray. But X-rays themselves can cause the tissue to become cancerous. In this Catch-22 situation many women are put off the idea of having their breasts monitored for cancer.

Soon, however, this dilemma may be a thing of the past if an X-ray-less mammogram from an American company, Somanetics, takes off. Using harmless light rays, the INVOS 2100 System measures the chemical make-up of the breast tissue. The 'likeliness' of an abnormal growth developing is indicated by a risk factor called the INVOS Value, which is available immediately. The system doesn't actually diagnose the situation – it determines the risk. 'It measures the biochemistry of the breast to determine the likelihood of breast cancer developing or already existing,' states David Weaver of Somanetics.

The new procedure is painless, harmless, low-cost and takes only five minutes. It can be performed as often as needed without causing any harm to your body. And as for accuracy, when the system was originally tested it detected 13 per cent of cancers missed by conventional X-ray tests.

STRESS GUM

●●● ✓ CONVENIENCE ✓ HEALTHIER

Stress gum is used by the Japanese as a quick and easy way of determining the state of their health and nerves. The gum reacts with the body's pH level to indicate the level of stress. After a few minutes of chewing, the user checks the colour of the gum. Pink is a sign of health; green is a signal that the user is suffering from unhealthy stress.

Right now, the manufacturer, S.B. Shokauhin Inc. of Tokyo, has no plans to market the product outside Japan. However, with everybody increasingly interested in monitoring their own health, stress gum is a natural for a wider market.

NON-INVASIVE HOME GLUCOSE TEST FOR DIABETICS

●●○ ✓ CONVENIENCE

Diabetes is an all-too-common illness among both adults and children. And while it can be controlled, it takes constant monitoring, insulin shots and a special diet. To make life easier for diabetics, a new device is being developed that can measure glucose levels in the bloodstream simply by being held against a person's skin.

And this may be just the beginning. Using the same technology, similar devices should be able to measure other chemicals in the blood, such as cholesterol and drugs.

The glucose tester has been developed by a Pennsylvania company in conjunction with a children's hospital in Toronto and a research institute in Columbus, Ohio. The device works by emitting a flow of energy directly into the patient's body tissue. It then registers numerically how the energy flow interacts with the tissue, indicating the glucose level in the body.

A large model for use in hospitals is already in existence. The research team hopes to reduce the device via briefcase-sized to hand-held.

FEVER CHECK

●●○   ✓ CONVENIENCE

Anyone who has ever sat up all night at the bedside of a child with a high temperature will appreciate this little monitoring device. Harry Bloch, a Pennsylvania banker and father of three, was on the horns of a dilemma when he woke to find his daughter's temperature had shot to an astronomical 105° as she slept. 'My wife and I didn't want to keep waking her every hour to take her temperature, but we knew we had to track the situation.'

Bloch felt there had to be some better way, so he invented a monitoring system that constantly tracks body temperature through sensors sewn inside the armpits of a pyjama top. The temperature can then be displayed in several ways. First, there's the illuminated display on each pyjama sleeve – visible in the dark. Alternatively, using the familiar wireless intercom principle, a beep can be sent to a unit in the parents' bedroom or an alarm can be activated if the temperature rises above a pre-set level.

Bloch, who has patented his device as the Infant Temperature Measuring Apparatus and Methods, feels it has uses beyond child care. He sees other people in need of constant monitoring – geriatric patients, for instance – benefiting from his invention, which is 'more sanitary and less intrusive than an oral thermometer'.

FLEXIBLE BONE CLAMP

●●● ✓ ENHANCEMENT

Doctors in West Germany and Italy are saving lives and improving the quality of life for others by using a new kind of clamp on fractured bones. Developed in Italy, it allows the bones to move fractionally while simultaneously holding them in the correct place.

The new device, named after its manufacturers, Orthofix, is the result of researchers realizing that broken bones heal best if allowed slight movement; this stimulates the growth of new bone. Previously bones had always been held rigid, which can cause such stress that new fractures occur.

A broken bone bleeds. The blood clot which soon forms is entered by cells which create a structure on which new bone can grow. A substance known as callus then begins to form on the broken ends of the bone. It will, however, disappear if the bones are held rigid. The fracture will seem to have healed, but the bone will still be weak and may fracture again. This undesirable situation is what Orthofix pre-empts.

It works like this. The surgeon attaches the device to the bone with four screws. The central section is telescopic, with a ball joint at each end, which enables the broken limb to be manipulated into the optimum position for healing. The central section is then locked until callus has formed. When it is unlocked, the bone is allowed slight vertical movement, which encourages the growth of new bone.

Orthofix is also working very successfully in bone-lengthening operations – for those with one leg shorter than the other, or for people suffering from dwarfism. In these cases the bone has to be fractured artificially; a screw in the central section of the device can then be regularly turned, and callus forms in the hole. When the bone has reached the desired length the telescopic section is locked and the patient begins to walk, thus encouraging still further the growth of callus.

PREGAPHONE

|●●●| |✓| WACKY

The Pregaphone is one of those wonderful ideas that started out as a joke and turned into a business. The device – an over-sized stethoscope in bright yellow with a mouthpiece at one end and a funnel at the other – is used to communicate with unborn children.

Dawn Hodson, a California business consultant, gave it to a friend as a present. Now she runs Pregaphone Inc.

'Research has shown that at six months children hear and learn in the womb,' she explained. And according to a study conducted by a pre-natal specialist, Dr Rene Van De Carr, children who were spoken to in the womb scored significantly higher than other children in tests dealing with early speech and use of compound words. 'There is also evidence that the bonding between child and parent is stronger when there is pre-natal communication,' says Ms Hodson.

SONIC PAINKILLER

HEALTHIER

After prevention and monitoring, we come to dealing with illness and disease that already exist. Pain is one of the great problems in treating disease and injury; and while there are very many effective painkilling drugs, they can often reduce the patient to a state of torpor and have unpleasant side-effects.

Throughout medical history some of the most intriguing discoveries have actually been accidents – penicillin is an example. Now a Sicilian-born physical chemist with two hundred inventions to his credit may have serendipitously hit on something that can relieve the intense, crippling pain of arthritis.

Dr Alphonse Di Mino was doing some research with electronic resistors. To test one of his theories he built a device called a Sonotron, which combined low-frequency radio and audio waves. When the Sonotron is on, it throws off a purple neon spark and generates a good deal of heat.

On a whim, Dr Di Mino decided to treat the arthritic pain in his hand using the spark as a sort of electronic heating pad. To his amazement, the pain stopped instantly. Further carefully monitored trials with other arthritis sufferers produced similar results. After several years of clinical testing in universities, researchers still aren't sure how or why it works, but their results indicate that the mysterious purple spark can ease lameness in horses as well as reducing arthritic pain in humans.

A model similar in size and shape to a stereo speaker is being manufactured for use in doctors' surgeries and hospitals. Dr Di Mino says that a smaller, less expensive model is under development for home use.

MICROWAVE PAINKILLER

ENHANCEMENT

We're all familiar with the techniques of acupuncture. It can relieve pain by inserting needles into the flesh, often at some distance from the painful area. Now the needles could be replaced by microwaves similar to those used in domestic ovens but at much higher frequency.

Dr Andre van der Vorst is undertaking research at the Catholic University of Louvain in Belgium. At these high frequencies, he says, microwaves can be generated on chips of a semiconductor material called gallium arsenide. A microwave chip and a battery to power it could be implanted under the skin at an acupuncture point.

They've been using this treatment in China for years – it combines acupuncture with another ancient medical technique, a heat treatment called moxa – but no one yet knows why it works. Dr van der Vorst and a Chinese colleague believe that the brain muddles up the acupuncture pain signals with the real pain signals, and is deceived into suppressing both.

SMART PILL BOTTLE

SAFETY

Let's face it, nobody likes taking medicine. In fact, many people don't take it – or they don't take it according to their doctor's instructions.

'Nearly one in every three patients misuses his drugs in a way that impacts on its effectiveness,' says Keith Mullowney, president of Aprex Corporation, an American company that has developed a solution to this problem. To put this more dramatically, if you have been prescribed pills to reduce the risk of a heart attack and for some reason stop taking them, the sudden interruption to your routine may actually trigger a heart attack.

Aprex's remedy is the 'smart' drug container. It is really just a special cap that will fit a standard pill bottle. The clever part is that it records the time and date every time the container is opened and closed.

The cap was originally designed for use by people participating in a new drugs testing programme in the USA. When each participant had finished the pills, he or she returned the cap to Aprex who computer-analysed the information in it and then prepared a report on when and how often the medication was taken.

If this kind of pill bottle were available to doctors, they would be in a much better position to see the effectiveness of the medicines they were prescribing. Ideally, patients would return the caps of used bottles to their GPs, who would read the records directly from a computer terminal on their desks. Clearly the technology is there; in Britain we would also need sufficient funds for the National Health Service to afford such an improvement.

Looking even further ahead, Aprex foresees caps with buzzers and alarms and liquid crystal displays. In this way patients could be alerted to take their drugs at the right time, and be given instructions about what to do if they had missed a dosage.

ELECTRONIC BANDAGE

ENHANCEMENT

Here's a British idea that will lend nature a hand by accelerating the body's natural healing processes. Surgeon Richard Bentall has created a disposable Electronic Bandage.

Actually a mini-electric system, the heart of which is about the size of a wristwatch, it speeds the healing process by pulsing electromagnetic radio waves into the cells of a wound. The Electronic Bandage can heal up to 30 per cent faster than normal, reduces swelling and leaves less scar tissue.

As Richard Bentall explains, an injured cell is like a rechargeable battery that has gone flat. The device, manufactured by an American company, recharges the battery, so to speak, by introducing an electronic charge. It can help internal wounds as well as external ones, and is normally left on for two to fourteen days after surgery.

The system, which is expected to receive US government approval very soon, will at first be available on prescription only and in hospitals. But its inventor is optimistic that eventually there will be a cheaper, over-the-counter version as production increases and costs go down.

ELECTRONIC DEODORANT

●●● ✓ CONVENIENCE

The prevention of body odour straddles the borderline between cosmetics and health care. But profuse sweating is unpleasant and irritating, and since a quarter of the population suffer from this complaint Drionic, an electronic sweat control device, may well be considered to improve their physical and psychological wellbeing.

An initial series of twelve or so self-administered treatments will result in six weeks of sweat-free bliss. After that, further six-week periods of protection will only need two to three treatments.

The process is called iontophoresis, and it stops emotion-related heavy perspiration by causing hyperkeratotic (thickening of a certain layer of the skin) plugs to form in sweat ducts. Worried about what happens when you block up the body's natural outlets? Studies have shown that when excess sweating of the hands, feet and underarms is stopped, there is no effect on the body's ability to control its temperature.

SUNBURN PROTECTION METER

●●● ✓ CONVENIENCE

Lying on a sun-soaked beach can be a wonderful relaxation – but it can also be very dangerous. And all those different-factor filters and lotions can be quite confusing, especially since on some days the undesirable ultra-violet rays are stronger than on others. Soon, a modestly priced device the size of a pocket calculator will take away the guessing and protect you from painful sunburn, ruined holidays and severe skin damage.

It's called the Sunsor UV Meter. Point it at the sun, and it will measure the strength of the ultra-violet rays on a scale from 1 to 120. Its findings will be displayed on a LCD screen. If you turn the meter over, a chart on the back will tell you in minutes how long it will be before you start to burn, assuming that you have average, unprotected, untanned skin.

KISS MOISTURIZER

●●● ✓ WACKY

At last, relief from the turn-off of a dry kiss! The key is something called a microsponge, a microscopic, synthetic sphere that can be programmed to release substances in response to pressure, time or temperature. In other words, when you pucker, your lips get moist. Microsponges are the brainchild of Advanced Polymer Systems of California, and the process, called Command Release, has been leased to a New York-based cosmetics firm. The result will be the world's first remoisturizing lipstick.

And in case you thought this was just a curious bit of trivia, the process is a highly scientific one and has applications far beyond lipstick and a good kiss. For microsponges can release not only cosmetic agents, but also pharmaceutical ones. They are sure to become a valuable tool for the health care professions.

Soon to be available at your chemist, the lipstick not only will remoisturize but will renew its colour all day long when you press your lips together.

LIVING SKIN EQUIVALENT

●●○ ✓ ENHANCEMENT

No guarantees exist yet, but the chances are that in the future we will be able to patch up our skin just like an old pair of jeans! A former Massachusetts Institute of Technology biology professor, Dr Eugene Bell, is producing living skin.

Organogenesis, Dr Bell's company, is not just growing skin cells; it is keeping the cells together in a total skin-like structure. 'What makes us unique', says Doug Billings of Organogenesis, is that we have a patented process that allows us to create skin "structure" from those cells, thereby making full-thickness skin.'

Reconstructive Tissue Filler (RCTF), one kind of skin that the company grows, is a fatty substance sometimes called 'deep tissue'. It could be used in plastic surgery to make certain parts of the body fleshier. But this is many years away, since trials have not yet been started.

The company is much further advanced with a product it calls Living Skin Equivalent (LSE), which uses the same cell culture technology as RCTF. It is grown from cells in the outer layer of the skin, the epidermis, and from cells in the inner layer, the dermis.

So far LSE and RCTF have no colour or smell and are unable to grow hair. The company is working on pigmentation of LSE so that it can be used to replace scar tissue with a natural look. Whatever the outcome, news of Dr Bell's work will surely be welcomed by plastic surgeons and accident victims.

SUPER PORE CLEANER

●●● ✓ ENHANCEMENT

From the dermatologist who brought you Retin-A comes a new cosmetic skin treatment that cleans pores more completely than anything else currently on the market. It's like having a professional come in and clean the carpets, or having the exterior of a building cleaned. The difference between the before and the after is amazing.

Retin-A, you will recall, is that wonder acne treatment that millions believe reduces facial wrinkles. Dr Albert Kligman's new discovery may not cause quite the same sensation, but for those who suffer the indignity of blackheads and the accompanying enlarged pores, a ten-minute treatment should bring relief.

What makes it so super? A substance – don't be put off – similar to what is found in glue! A liquid containing the ingredient is applied to the skin – for the time being, only trained professionals carry this out. When it dries it is peeled off, taking with it dirt, oil and bacteria from the pores.

According to Kligman it works best on blackheads, removing the debris that makes them look enlarged. It is effective before or after you get a spot, but not during. He claims his product is 100 per cent better than existing face packs aimed at dealing with the problem. 'A little discomfort' and temporary reddening are sometimes experienced by the patient, but this is soon forgotten when the skin becomes 'healthy, clean and smooth'.

WRINKLE-REDUCING PILLOW

ENHANCEMENT

Everyone knows that too much sun, too much worrying and too much smiling can cause wrinkles, but now experts claim that even sleeping can damage the skin.

Ordinary pillows – whether soft and fluffy or hard as a rock – put pressure on the face, thus stretching the skin tissue and eventually causing wrinkles. The Goodbye Sleeplines pillow, the current brainchild of an American company called La Jeunesse, is designed to alleviate that pressure, while providing proper support for the head. Advertised as an alternative to collagen treatments, the pillow helps the skin maintain its tone and elasticity.

The 16 x 23 inch rectangular pillow has a hollow centre cut out in a shape that improves the circulation of blood to the face. Since most people sleep on their sides, they put pressure on their faces, says Carol Bianco, executive vice president of the Sleeplines Division of La Jeunesse. 'You're putting in creases every night, and they get more and more pronounced. It's almost like ironing in the same wrinkles every time you lie down.'

The pillow was invented by Kerry Lake, an interior designer, who noticed that she never looked her best for breakfast meetings. So she designed a foam support that didn't press against her skin. In a short time, she was looking a lot better early in the day.

The pillow is also ideal for anyone recovering from facial or dental surgery, Carol Bianco says. 'There's no pressure on the face, so healing is accelerated.' Made of polyurethane foam, the pillow is also recommended for people who need optimal back and neck support, since it puts the spine in correct alignment for all sleeping positions.

SNORE-REDUCING PILLOW

ENVIRONMENT-FRIENDLY

No one likes to suffer insomnia while their companion snores the night away. But Australian grandmother Judy Challen turned her sleepless nights to good use by sketching her husband's head and neck postures. By day, she used a kitchen knife to fashion a foam pillow based on these sketches, in the hope of creating a pillow that would alleviate the problem. Her work finally yielded a shape that tilted her husband's head back, instead of forward, and allowed him to breathe more quietly. With the rudimentary pillow in hand, Mrs Challen approached a company that manufactures foam cushions and pillows, and they agreed to mass-produce her invention in a flame-retardant and non-allergenic material that retains its shape.

The snore-reducing pillow is aimed at people with 'positional' snoring problems, rather than those with physical breathing difficulties. It differs from an ordinary pillow because its centre-ridge construction supports the head from any sleeping position and eliminates air blockage in the larynx and upper airway – the most common cause of snoring.

By the way, Judy Challen – and her husband – both sleep soundly now.

ELECTRIFIED BIRTH CONTROL

●●○   ✓ CONVENIENCE

While driving to work one day, a New York gynaecologist, Dr Stephen Kaali was thinking about the battery that needed changing in his watch. His thoughts wandered, and Dr Kaali, a family planning researcher, suddenly wondered if a small battery inserted in a diaphragm might not kill sperm.

He has now taken out a patent on just such a device. The battery is cylindrical, a quarter of an inch long and the width of the stick on a cotton bud. Built into the rim of the diaphragm, it sends its signal out across the dome and kills sperm by zapping them when they cross the low-level electrical field.

The battery, which lasts six months, produces only 50 microamps at 2.8 volts – the identical current created by a pacemaker and even less than that of a battery in an electric watch. 'No one would feel anything from the current, except the sperm,' Dr Kaali explained. And so far, he adds, it has proven 100 per cent effective in lab studies and animal experiments.

The electrical birth control device has also shown that it can kill bacteria and fungi as well as sperm. According to Dr Kaali, it should cure yeast infections, and research is being conducted to determine if it could even kill the AIDS virus at the moment it is passed on.

TIME-RELEASE BIRTH CONTROL

CONVENIENCE

The trouble with birth control pills is that you have to remember to take them. Now, thanks to research conducted by the Population Council, a non-profit-making international organization, there's a new method that's more convenient and needs no reminders.

Researchers have developed a time-release implant method using six thin, match-size, hormone-filled capsules. The implants are inserted surgically, in a fan shape, on the inside of a woman's upper arm. They're flexible and comfortable. Suitable doses of a progestin hormone are released at a slow, steady rate, preventing pregnancy within hours of insertion and lasting up to five years. So far, studies show a less than 1 per cent failure rate, per year.

Norplant, the registered trademark, is totally reversible and can be removed surgically at any time. It has already been approved in twelve countries, including China, Sweden and Finland, and clinical trials are under way in forty others. In the West this kind of birth control is merely convenient; in overpopulated regions of the Third World, where women are not able to understand the necessity of taking their pills on a regular basis, it could be a miracle worker.

IMPOTENCE PILLS

CONVENIENCE

Every year impotence – or erectile insufficiency, to use the medical jargon – leaves millions of men feeling sexually and emotionally inadequate. To date, injections and psychological treatment have been the methods used to combat this most private of problems – not always with success. And if the condition is traced to purely physical causes, doctors often prescribe an injection to be administered just before intercourse. Naturally this limits spontaneity and can be, to say the least, awkward. But now Dr Grant Gwinup, a professor of medicine at the University of California, is offering new hope to these unhappy men – and their partners.

It's a pill containing a substance called phentolamine,

and in initial tests it helped eight out of sixteen impotent men maintain erections long enough to complete intercourse. The pill is safer and easier to take than the injection, which can cause infection in rare instances or, more commonly, a condition known as priapism – continuous, non-sexual erection. But there is one drawback: when injected, the drug goes directly into the bloodstream, becoming effective in minutes; the oral dose takes up to an hour to work.

The international drug firm Ciba-Geigy used to manufacture phentolamine for use in counteracting hypertension. But unfortunately when an improved product came along the company ceased production. 'Ciba needs to be convinced that there's a market for this,' says Dr Gwinup. He would like anyone interested in seeing phentolamine made available again to write to the Ciba-Geigy office in their particular country.

RADON EXTRACTOR

✓ SAFETY

The carcinogenic gas radon was much in the news in the late eighties and early nineties, as the West Country was found to have unacceptably high levels of the substance in the ground. America, too, has identified this problem, which comes from decaying uranium in the soil and seeps up from the ground into basements. It is when trapped like this that it becomes dangerous.

To reduce the risk a Massachusetts company has developed RAdsorb-222, a self-contained radon removal

HEALTH MATTERS IN THE FUTURE

unit which removes 98 per cent of the radon in the air and requires only a small amount of electricity to operate. The 4 foot high box works by absorbing radon-contaminated air into charcoal filters, where the air is purified and the radon is trapped. The radioactive gas is then vented to the outdoors, where it is rendered harmless.

PHONE SHIELDS
●●● ✓ HEALTHIER

You don't have to be Howard Hughes with his fear of contamination to feel revolted by the average public telephone box. A Massachusetts entrepreneur, tired of having to hold the receiver several inches away from his ear in airports and the like, has come up with a product to deal with the problem. The OliverShield, brainchild of Anthony Oliver, is a paper guard that sticks to the mouth- and earpiece of the receiver, thus blocking out nasty germs. It uses a sticking plaster-type, non-toxic adhesive and can easily be peeled off the receiver when you're ready to hang up.

VILEST TASTE
●●● ✓ SAFETY

Described as 'the bitterest substance in the world' by its manufacturer, Atomergic Chematals Corporation, denatonium saccharide is a non-toxic chemical compound that, even when diluted in the ratio of one part per ten million, retains its distinctive, revolting flavour. If you're wondering what was the point in developing such a product, the answer is as a deterrent and safety device.

The main use of this substance will be as an additive to certain poisonous compounds, including household chemicals to warn people (especially children) to stay away from them. It's similar in function to the nasty smell put in otherwise odourless domestic gas – so we know when the pilot light's blown out or there's a gas leak.

The company has also formulated an animal, rodent and bird repellent using this ingredient. Called Ropel, when spread on dustbin rubbish it deters even the hungriest foraging animal for quite some time. 'To taste Ropel is to know how it works,' says Mel Hollander, one of its inventors. 'It lingers for several hours, and only a mouthwash or a stiff whisky can make it go away.'

DIAGNOSTIC LOO

WACKY

Here's the answer for all those hypochondriacs on your Christmas list. It will help put their medical fears aside by providing instant read-outs of their health whenever they feel the urge.

Toto Ltd, Japan's largest manufacturer of sanitary ware, is interested in marketing an 'intelligent' loo that takes your temperature and your blood pressure, analyzes your urine and faeces, and weighs you whenever you use it. The great advantage of this new bathroom fixture will be that a patient's health condition can be monitored by a doctor without the patient having to leave home – or even the bathroom! For this most extraordinary of loos will be able to transmit its readings via telecommunications equipment to a doctor's surgery.

NIT, the Japanese telephone and telegraph company, is in on the development of this product, as is Japan's largest manufacturer of electronic clinical thermometers, Omron Tateisi, which also manufactures electronic systems for transferring money.

FREEZING HUMANS

☑ TIME-SAVING   ☑ WACKY

This is definitely one for the end of the century, or perhaps even later. But we're talking 'when' and not 'if' – freezing humans for the purpose of preservation is a medical advance that will almost certainly become reality some day. At present, money seems to be the only brake on this exciting new development, which is known as cryonics.

'The technology is not far away,' says Hal Sternberg, a leading cryonics researcher at the University of California at Berkeley. 'But some say, "It's far-fetched, so why invest?" In reality, cryonics technology may be simple compared to the biological realities of curing cancer or AIDS.' And, of course, the main reason for freezing is to keep people 'alive' until cures for such deadly diseases and for ageing can be found.

This is what the scientists have achieved so far. A beagle named Miles (after Woody Allen's character in Sleeper) was anaesthetized, laid on a bed of ice until his temperature fell to 68° Fahrenheit, emptied of his own blood, filled up with a synthetic blood that wouldn't clot in the cold, and then almost frozen. For the hour he spent at or below 50° Miles was clinically dead. 'Thawed' and reinjected with his own blood, Miles 'came back' and proceeded to lead a normal life, 'in perfect health', assured Sternberg.

The next step is monkeys, and ultimately people. But there are still obstacles to overcome. Experiments are now being conducted to determine how to minimize the microscopic damage to tissues that takes place during the freezing process. Although there are people who have great reservations about cryonics, others view it with tremendous hope.

5
FLY DRIVE INTO THE FUTURE

Collision avoidance system
Car video navigation system
Car satellite navigation
Turbine car engines
Plastic car engines
Two-cycle car engines
Drive by wire
Heads Up Display
Night vision display screen
Light-sensitive car windows and mirrors
Rear-View TV monitor system
Automatic tyre check and fill
Never-flat spare tyre
Seat bicycle pump
Four-wheel steering
Smart suspension
Remote control car starter
Flying car
Levitation vehicle
In-flight entertainment system

COLLISION AVOIDANCE SYSTEM

SAFETY

The idea of a collision avoidance system is something that's bounced around the motor industry for years. After all, planes and ships use radar to avoid collisions, so why not cars? The problem with cars is that they travel so close together that there's little time to react once you've been alerted to danger.

Now Nissan, the Japanese manufacturer, is working on a system that does more than warn you of an impending crash, a jaywalking pedestrian or some obstacle on the road. It acts. Nissan's system projects a pulsating laser beam in front of the car. The beam is reflected by any object in its path to an optical head mounted on the front bumper. A computer then calculates the distance by measuring the time required for the beam to return to the sensor. The computer almost instantaneously measures the hazard level. If you're travelling too fast, it will instruct the accelerator to ease up. And if you're about to hit something or someone, it will slam on the brakes.

There's still work to be done on this system. But experts believe the odds are better than even that this type of collision avoidance system will be standard equipment by the year 2000.

CAR VIDEO NAVIGATION SYSTEM
CONVENIENCE

If we can land men on the moon and planes can find their way to Heathrow on the foggiest of nights, how come we're still using antiquated paper maps to find our way on the roads? It's not necessary any more. Electronics has taken us far beyond: video display route charting is here, in Britain it takes the form of the road guidance system called Autoguide.

Details will vary as the project develops, but basically it works like this. Cars fitted with Autoguide will have an infra-red receiver and transmitter fitted behind the rear view mirror. Beacons will transmit details of the road network in the locality. The receiver is linked to a central processor which at present is large, but miniaturization is on the cards.

On setting out, the driver keys in the map reference of the destination, which is displayed on one of two dashboard screens. An arrow on the second screen then points to the destination. When the car passes a beacon the driver will receive further guidance, which includes on-screen arrows for turns, and spoken instructions from an electronic voice. The system can respond quickly to a change of situation. Problems encountered, such as traffic jams, are fed into a central computer which may then adjust instructions issued to other cars wanting the same destination.

CAR SATELLITE NAVIGATION

●○○   ✓ CONVENIENCE

The video navigation system (see previous entry), soon to be introduced, will be replaced at the end of the century by satellite navigation. This amazing system will be able to pinpoint your car's position anywhere in the world, alert you to a traffic jam ahead and show you all the possible alternative routes, warn you of approaching bad weather, and so on. Your car will be bouncing signals off satellites in space just like the most advanced communications systems.

Nissan of Japan is spearheading the technology with its Satellite Drive Information device. Your position will be shown on your car's computer display screen, using Nissan's Global Positioning System. Forget video road maps and memory banks. The picture on your screen will be the real thing transmitted via satellite. The display screen will even tell you where to make turns and will also measure distances – to your motorway exit, say, or to your final destination.

Satellite navigation is going to make it difficult for even those of us with no sense of direction at all to get lost!

TURBINE CAR ENGINES

●●○   ✓ ENVIRONMENT-FRIENDLY

Turbine engines are coming to cars. That was the word from the Chrysler Corporation twenty-five years ago. If you're still waiting, don't give up – but don't rush to your local car dealer either. They're coming – but still not yet.

Similar to the kind used in jet aircraft, car turbine engines are light. They have fewer parts and therefore fewer mechanical problems. They can use a variety of fuels, and they last a very long time. The drawback is that, while the costly, specialized materials make the turbine engine OK for use in a plane, it's out of the price range for a family car, despite the obvious advantages.

The bottom line? A revolutionary breakthrough in ceramics and other high-temperature materials could make the turbine and the motor car familiar bedfellows by the turn of the century. Otherwise, turbine engines might find their way into only top-of-the-market sports cars.

PLASTIC CAR ENGINES

ENHANCEMENT

And here's another car designer's dream: to replace an engine's complex, heavyweight iron castings with a single extrusion of plastic. At the dawn of the 1990s that fantasy is close to reality, thanks to fibreglass-reinforced composites.

The major car manufacturers have already seen running prototypes of plastic engines that show real promise. They're light and they're strong, and according to their inventor, Matty Holtzberg of New Jersey, they're destined to be the engines of the future.

Still, the giants of the motor industry are slow to

jump on a new technology when it means a major change in production. As a result, we're likely to see advanced plastic composites used inside engines – connecting rods, pushrods and the like – before the entire casting goes plastic.

But don't be surprised if new ultralight economy cars appear before the year 2000, with hearts of plastic under their bonnets. The technology is here now.

TWO-CYCLE CAR ENGINES

✓ MONEY-SAVING   ✓ ENVIRONMENT-FRIENDLY

Car makers all over the world are taking a hard look at a two-cycle engine developed by Australia's Orbital Engine Co.

The incentives are there. With far fewer pieces inside, the engine is inexpensive to build. And it's lighter, too, which means it consumes less petrol. Two-cycle engines operate without valves or camshafts, producing power with each complete turn of the crankshaft. Standard four-cycle car engines generate power on just half of their revolutions. The result is that Orbital's 90 lb engine can improve on the output of four-cycle engines even though it's one-third the weight.

Additionally, this more efficient combustion process makes it easier for the car to meet Australian exhaust emission regulations, according to the inventor, Ralph Sarich. His engine offers considerable refinements over the old two-cycle engines used in outboard motors and in the Swedish Saab cars of the early sixties. There's no need to mix oil and petrol in advance, and modern noise controls have almost eliminated the raspy ring-a-ding sound of those earlier engines.

According to the experts, how soon two-cycle engines will appear in car showrooms depends on the politics of royalties, patent rights and the willingness (or unwillingness) of manufacturers to retool their production lines. If Detroit dallies, says one motoring writer, look for Tokyo to leap at the chance to go two-cycle.

DRIVE BY WIRE

●●○ ✓ ENVIRONMENT-FRIENDLY

It sounds a bit scary: a car's steering and speed controlled by sensors connected only by thin wires rather than by all those heavy mechanical links still found in the cars of the eighties. But drive by wire is almost certainly the way of the future.

Basically, all the bulky tubes and rods and other mechanical bits and pieces on present-day cars will be thrown out in favour of a system that operates on sleek little electronic sensors connected by skinny wires. The sensors will detect when you put your foot on the accelerator or the brake, or which way you turn the wheel and how much. They will transmit what they 'sense' to a central system, which will then issue the right command.

There are many advantages. Driving precision is increased, for starters. And with all the bulky metal parts replaced by thin wires, the car will not only be lighter but also quieter, since the gaps in the car's bulkhead will only have to be big enough to pass a small wire through.

The only car currently available that runs on drive-by-wire linkage is the BMW 750L, and it's certainly not at the bottom of the range! But although it's all new to cars, planes have been flying by wire for quite some time. So if this talk about little wires operating your car still worries you, remember that the system has proved safe for air travel. Soon it will be on your car.

HEADS UP DISPLAY

●●● ✓ SAFETY

From the time the first quaint motor cars appeared over a century ago, the task of keeping the driver informed of operating conditions such as speed and level of fuel has been vital. Yet presenting that information has always required the driver to turn his attention from the road ahead and search the dashboard for the appropriate gauge or meter. Now, thanks to technology adapted from the leading-edge information display systems of military fighter jets, that undesirable state of affairs is no more.

With Heads Up Display, the windscreen is covered with a thin layer of transparent material bearing a holographic image. A device mounted on the dashboard activates the hologram by illuminating it with light of a specific wavelength, and as a result an image is projected 8 feet ahead of the vehicle into the driver's field of vision.

The system, developed by Hughes Aircraft for General Motors, shows speed, low fuel level, full beam headlight indicator and turn signal. The obvious advantage is that the driver no longer has to take his or her eyes off the road to check the instruments. But more than that, eyestrain will be avoided by not having to adjust for the change in focus distances and outside/inside light levels.

NIGHT VISION DISPLAY SCREEN
●○○ ✓ SAFETY

Here's another helpful bit of technology that will come to us courtesy of the military. It's a night vision system that will let drivers see through darkness, fog and rain far better than with the most powerful headlights.

At present being developed by General Motors, the system uses a special receiver sensitive to infra-red rays that projects on to a dashboard-mounted TV screen an image of what's in front of the car. The picture is actually formed by heat values emitted from objects in the car's path.

GM has a prototype infra-red night vision system, but it's large and must be mounted on the car roof. Furthermore, the receiver has to be cooled by liquid nitrogen. Producing a smaller unit and solving the heat problem are the challenges that must be overcome before this system makes it to the showrooms.

LIGHT-SENSITIVE CAR WINDOWS AND MIRRORS

SAFETY

The idea of car windows which adjust their tinting to the brightness level outside seems straightforward enough. After all, haven't those sunglasses been available for years?

Ford Motor Co.'s Glass Division has taken the concept a step further. It's called Switchable Privacy Glass, and uses LCD technology just like the display in your digital watch. Basically it adjusts the transparency of the glass from fully transparent to nearly opaque.

A slight variation will be used in rear-view mirrors. Depending on the illumination falling on the mirror from the headlights of cars behind, the mirrors will vary their reflection to offer glare-free rear vision. Additionally, an electro-luminescent halo surrounding the mirrors will help the driver locate them quickly in the dark.

Not surprisingly, General Motors is working on the same idea, though their technology involves electrochromic glass, which darkens or lightens when an electric sensor is fed through it. 'I expect to see glass roofs, the kind where you dial in opaque or clear depending on how hot the sun is,' says Charles M. Jordan, head of corporate design at GM. 'It's likely that the whole upper part of the car will be glass.'

REAR-VIEW TV MONITOR SYSTEM

✓ SAFETY

Maybe you've seen this system on those small airport buses that take you to your hire car. A video camera is mounted on the rear of the vehicle, and a small black and white TV monitor on the dashboard. The result is a clear, wide view with no blind spots.

The device should prevent accidents and save lives, but it will be the mid-nineties before it goes on sale to the general public. The Automotive WatchCam is manufactured by Sony – but so far it's only being sold in the lorry, bus, boat and recreational vehicle markets. It is particularly helpful to drivers trying to manoeuvre a vehicle into a tight parking spot. As you can imagine, the bigger the vehicle, the more helpful the WatchCam, as looming back ends can obstruct the driver's view for as much as 30 feet.

The systems are small, as space is at a premium in most vehicles. The monitors are either 4.4 or 5.5 inches; the larger unit comes with day/night switches for picture adjustment. The camera comes in a water-resistant housing and its cable can withstand temperatures from –22° Fahrenheit to +167°.

Although cars have only a small blind spot, Sony believes the monitor system will still be a viable product for car drivers. The company believes it will instil confidence.

AUTOMATIC TYRE CHECK AND FILL

☑ SAFETY ☑ ENVIRONMENT-FRIENDLY ☑ MONEY-SAVING

If you've ever had a blow-out on the motorway, you're really going to appreciate this latest bit of motor industry technology. The device, called entireControl and manufactured by techni Guidance of California, uses sensors mounted on each wheel to detect any change in tyre pressure. This information is then relayed to the car's central computer. 'If the pressure decreases to 15 lb per square inch, a CHECK TYRE warning will begin to flash on the dash and an audible beeper will warn you to get the car off the road,' explains inventor Shrikant Gandhi.

This system is already available on a few luxury cars. But techni Guidance's latest advance takes this helpful device one step further. From a reservoir mounted on the inside of the wheel rim, carbon dioxide can be pumped into the tyre. The system monitors the tyre pressure in all four tyres and automatically adjusts the pressure to maintain it at the pre-set, driver-programmed level. The result is better road handling, improved fuel economy and peace of mind. Guaranteed proper pressures save tyres – and they save lives too.

NEVER-FLAT SPARE TYRE

●●○ ✓ CONVENIENCE

Until the automatic system (see previous entry) gets here we'll all still need to carry a spare – and most of us have at some time been faced with an unusably flabby spare at the roadside because we forgot to pump it up at the garage. Here, then, is an innovation that seems long overdue: a tyre that won't go flat.

Uniroyal Goodrich has developed a spare made of polyurethane elastomer, a material that's both tough and flexible. It's these properties that give the 'no air' spare the ability to provide a smooth ride while at the same time resisting the kind of road hazards that make ordinary pneumatic tyres go flat.

The new tyre has a paddle wheel design that's light yet durable. Its rubber tread is similar to that of its 'air' counterpart and should be good for three thousand miles of driving. It's also 20 to 25 per cent lighter than a conventional spare and occupies 35 per cent less space in your boot, according to the manufacturers.

SEAT BICYCLE PUMP

●●○ ✓ SECURITY

And while we're on the subject of tyres, let's just slip away from cars for a moment to look at the humble bicycle. If you're a cyclist, you'll know the risks of leaving your bike somewhere with the pump visible – you'll probably never see it again. Now Gilmore Chappell, a mechanical engineer from Pennsylvania, has come up with a revolutionary new kind of pump.

What you do is replace the seat post (the bar that holds the seat and connects to the bike frame) with what Chappell has named the Seat Post Tyre pump. Attached to the pump and nestling under the seat is a storage bag containing a coiled hose. When one of your tyres need filling, you simply pull out the hose and connect it to the tyre's air valve.

Pumping it up is the neat part. Your seat serves as the handle – you just move it up and down to fill the tyre with air. When you've finished, the seat snaps back into place.

FOUR-WHEEL STEERING

✓ SAFETY ✓ CONVENIENCE

Of course four-wheel steering is already here. But so far it's a flashy novelty on sports models rather than an important safety and driving feature on the family car. By the mid-nineties, however, four-wheel steering should be as common as front-wheel drive.

The strength of four-wheel steering is manoeuvrability. When you're driving at low speed the back wheels turn about 5° in the opposite direction to the front wheels. It may not sound a lot, but it makes a huge difference when you're trying to park in a tight space. At high speeds, four-wheel steering greatly improves stability, particularly when making fast lane changes on motorways.

So four-wheel steering isn't just for speed merchants and car freaks. It's for every driver. Right now, only the high cost is holding it back. But greater mass production should bring the price down and put it on every car except the lowest-priced models.

SMART SUSPENSION

●●○   ✓ ENHANCEMENT   ✓ SAFETY

Before too long, cars will be coming with no springs attached. Shocked? Don't be, because cars won't have any shocks either!

Replacing coil springs and shock absorbers will be active suspensions – systems that adjust automatically to bumps and potholes. With hydraulic rams supporting the chassis, active suspensions will smooth the car's ride before the tyre can rise out of the pothole. Built-in sensors will let a central computer know exactly what is happening at each corner of the car. The driver can programme this system for a ride that's 'soft', yet still as responsive as a sports car. 'The old days of compromise have gone,' says Michael Kimberley, chief executive at the British sports and racing car company, Lotus Engineering, where the system originates. 'You can literally programme whatever ride and handling behaviour you want.'

At present, active suspensions are being used only in Grand Prix racing cars. The high cost is keeping them out of mass production for the time being. But the technology is here, and some day you'll have an active suspension in your car.

REMOTE CONTROL CAR STARTER

☑ CONVENIENCE

Whatever your car – be it a sleek state-of-the-art job with aerodynamic styling, or a family runabout that's seen better days – you've still got to start it in the mornings, in all weathers. When it's cold outside you start your car from as much as 400 feet away with Auto-Command Remote-Control Car Starter.

From inside your home or office, a touch of the transmitter button on your keyring will turn on your engine, as well as the heater, the heated rear window and the windscreen wipers. The starter consists of a small box hooked up inside your car with fifteen wires that serve as sensors to turn on the appropriate functions.

'The system actually monitors the vehicle,' says Mike Stern of Design Tech International Inc., the manufacturers. For example, if you start your car remotely but no key is inserted into the ignition within ten minutes, the car engine is turned off. Or if a thief tries to steal the car without the key, the engine turns off when the brake or accelerator are activated. It also turns off if, for instance, a child were to get in and take the car out of PARK without inserting the key.

FLYING CAR

●○○ ✓ WACKY

Is it a car?! Is it a bird?! Of all the weird and wonderful inventions in this book, this must be the wackiest.

Start with a standard Honda CRX; add wings, a tail, another engine; make a few modifications; and you have it. A car that actually flies! And when you're on the ground, the 34 foot wings fold up to form a 20 x 8 foot trailer that's hauled behind the car.

The Aerocar is the dream of one man: a former US Navy pilot and aero engineer called Molt Taylor, who's been beavering away at his brainchild for more than thirty years. His Mark 3 Aerocar has actually flown and is now in the Museum of Flight in Seattle. Mark 4 has received government approval to be sold in kit form.

As if a flying car isn't incredible enough, the invention uses most of its standard car equipment when airborne. Explains Taylor, 'The same controls fly and drive it, but the steering wheel moves back and forth to operate the elevators. And there are rubber pedals that lie on the floor when you drive and come into position when you're ready to fly.'

The plane's turbine engine uses jet fuel, while the car engine uses normal petrol. The Aerocar meets all the US government requirements for ordinary cars, including those for fireproofing, pollution control and safety.

Taylor, now in his seventies, is in need of more finance and doesn't yet know when fully manufactured Aerocars will be available for sale to the public. He admits that the government is hoping it won't be too soon. 'The problem', he explains, 'is how anyone is going to control thousands or even millions of flying cars!'

LEVITATION VEHICLE

●●● ✓ SAFETY ✓ WACKY

This too is the stuff of comic books and sci-fi magazines, and the dream of generations of little boys who loved machines.

The Moller 400 looks like a sleek cross between a sports car and a rocket ship. It's a car, a helicopter and a plane all in one. It seats four, takes off vertically, can do up to 400 miles an hour, hover low, land softly and be parked in your garage. And it's almost as easy to operate as a video game.

The inventor is Paul Moller. As with Molt Taylor, this is a vision he's held on to for a long time. While researching for a PhD at McGill University in Canada, and through fifteen years of teaching at the University of California, he worked to develop new types of aircraft. Now head of his own firm, Moller International, he is putting the final touches on his masterpiece, which he modestly calls 'an alternative to the family car'.

He has already tested the technology for the Moller 400 in his earlier model, the 200X, which looks like a flying saucer. It operated successfully on numerous flights – both by remote control and with a pilot on board. Now the Moller 400 is about ready for take-off. It's 6 feet high, 9½ feet wide and 18 feet long. It has an economy cruising of 225 miles per hour and does 15 miles to the gallon, and it's powered by eight 65 lb, 528 cc rotary engines. Each engine generates 150 horsepower, four times that of a typical aircraft engine.

These eight compact engines are encased in four ducts. With no exposed blades, the craft is much safer to manoeuvre on the ground than either a helicopter or a small plane. Indeed, the Moller 400 has been built with safety in mind. Three on-board computers check each other's work and can back one another up. They also provide the aircraft with a sophisticated collision avoidance system. At speeds above 125 mph, altitude can be maintained even if six of the eight engines should fail. And if all the engines should die, the Moller 400 will land with the aid of an emergency parachute. Also, its 5 foot stiletto nose will crumple to absorb shock.

While it may seem like the fulfilment of every commuter's fantasy to leave bumper-to-bumper traffic below, Moller believes that the craft's first application will be performing search-and-rescue missions in isolated areas. Still, there are a lot of childhood dreamers already putting their names in the order book. Michael Jackson and Richard Branson have expressed their interest, and by early 1990 over fifty people had paid their deposit and reserved a Moller 400.

Is this the way of the future? Peter Mustafa, agent for Moller in the UK, says 'The company could be as big as Boeing. In ten years' time, millions of these machines will be on the road and in the skies.' And if you think, like Molt Taylor in the previous entry, that that will create the same problems that we see in today's traffic jams and stacked incoming planes, Moller believes that it could all be ironed out by the development of a satellite-controlled navigation system.

IN-FLIGHT ENTERTAINMENT SYSTEM

✓ ENHANCEMENT

Those of us who can't afford our own plane, but have to travel on the ones belonging to the airlines, can at least look forward to some good entertainment in the nineties. Boeing and Sony have teamed up to create an in-flight information and entertainment system that rivals anything you might now have in your home.

Here's what the two companies have in mind: a 4 inch flat-panel display screen located on each seatback, allowing passengers to choose their own entertainment. The choices are movies, TV shows, video games or live viewing of the landing and take-off of your plane. Passengers will also be able to order food and drink and even duty-free goods through the system.

Other available information will include gate directories, the status of connecting flights and a moving route map. Service on board should improve as well. The individual consoles will be linked to a central cabin management system that will give flight attendants instant information on passenger needs, as well as keeping a food and drink inventory and a cabin maintenance report.

6
REVOLUTIONARY SHOPPING

Electronic supermarket
Supermarket self-checkout
Supermarket smart card
Safer tamper-proof packaging
Self-cooling can
Frozen beverage mug
Biodegradable plastic bags
Fresh fruit wrap
Freeze-dried compressed food
Noise-dried orange juice
Potato ice cream
High-fibre cakes
No-calorie sugar
Non-fattening fat
Cholesterol-free eggs
Carbonated milk
Self-stirring saucepan
Mini portable oven
Deodorant underwear
Ultrasonic 3-D clothes
Hot/cool fabric
Xerographic bedlinen

ELECTRONIC SUPERMARKET

✓ CONVENIENCE ✓ TIME-SAVING ✓ MONEY-SAVING

Supermarkets are about to change radically, and for the better. Coming soon is a system of electronic price tagging that's going to make your store more efficient, at the same time as it makes you a smarter shopper.

When you pull your trolley up to the baked beans, instead of card labels showing the price per can and per unit you'll find an easy-to-read LCD screen displaying this information. It can also print messages such as BEST VALUE or LOW CALORIE. Every one of the store's mini-screens is connected to the manager's main computer via low-power radio signals. Prices can be changed on one product or on all products just by keying in some numbers. And the new prices are instantly transmitted to the cashier's checkout scanner (which, of course, not all British supermarkets yet have!). So the price you see in the aisle is the one you'll pay at the checkout.

The company at the forefront of this technology is Telepanel Inc. of Ontario, which has already installed the system in several test supermarkets. 'Consumer response has been very, very positive,' says Telepanel.

Supermarkets love it too, because it reduces the workload and offers competitive flexibility. If, for example, it's discovered that prices of certain goods are being undercut by a neighbouring rival store, at the touch of a button the manager can reduce the prices on his own stock. Managers will also be able to affect the customer flow in their own stores. If Saturdays are busiest but Tuesdays are slack, special promotions, reductions and loss leaders on Tuesdays would probably attract some of those weekend shoppers.

SUPERMARKET SELF-CHECKOUT

✓ CONVENIENCE   ✓ MONEY-SAVING

What automatic machines did for banks, self-checkout machines will do for supermarkets. The idea was dreamed up by a man named David R. Humble – the inventor of security tags for garments – while waiting in a queue at a checkout. The person in front of him took the liberty of scanning a purchase by himself when the checkout assistant was distracted. The customer seemed to be delighted by the process of running the bar codes over the scanning device, and thus the idea was born that shoppers might like to check out their own groceries.

Humble has since developed a system called the Automatic Checkout Machine (ACM), that is already being used in a small number of pilot stores. More will surely follow.

Once each purchase has been scanned by the shopper, it goes down a conveyor belt with built-in security – just in case an item should sneak by unscanned. At the same time, the price and description of each item are displayed on a touch-sensitive screen. This screen has a TOUCH button for a running total of purchases – a welcome addition to supermarket shopping – as well as one for the final total. A human still packs up the groceries and puts them in a trolley to wheel to the car park. And if the shopper feels confused at any point, a

HELP button on the screen will summon assistance from store staff.

After all the purchases have been scanned, the shopper touches TOTAL on the screen to see what is owed and gets a printed bill. He or she then goes with the trolleyful of shopping to a manned pay booth.

SUPERMARKET SMART CARD

●●○ ✓ CONVENIENCE

Smart cards for the supermarket might be as popular in the future as automatic machines outside banks and building societies. They will debit and credit your account and can be used as cheque cards, if you wish.

But, most importantly, customers will accumulate points for coming back to the same supermarket chain. Once a smart card shows a certain number of points, its owner will be entitled to a prize – maybe some free items from the store, maybe a discount off future purchases, or maybe even a gift from a catalogue.

Supermarket managers and the manufacturers of the goods they stock can automatically gear promotions to customers who are most likely to take advantage of them. A card inserted into the screen at the checkout might reveal a long list of past baby item purchases. In this case the system might be programmed to respond with an automatic coupon for a new flavour of baby food, or money off the shopper's next purchase of her favourite brand of nappies.

All the information will be stored on the card as if it

were a computer disc. No longer will you have a kitchen drawer stuffed full of tatty coupons that you always forget to take to the shops with you!

SAFER TAMPER-PROOF PACKAGING

SAFETY

It's always alarming when someone with a point to make, or with a grudge against society, starts contaminating the food on supermarket shelves. In Britain in the late eighties it involved baby food, but some time this decade a much safer form of packaging should do away with mums' fears.

The new packaging is the invention of Dr Kim Krumhar, who came up with the idea while conducting research at the Massachusetts Institute of Technology. The object was to find a common, non-toxic substance for food and drug use that would cause a seal to change colour permanently when exposed to air, thus indicating that somebody had tampered with the container. Several years later Dr Krumhar, now with a company called Frito-Lay which is developing the product, is still perfecting his design. At present the sensor changes from blue to orange when exposed to air. What he wants is a more dramatic and meaningful colour change – such as from green to bright red.

SELF-COOLING CAN

●●● ✓ CONVENIENCE

For a few pence extra picnic lovers, warm beer haters and people with packed fridges will soon be able to buy drinks in self-cooling cans. Dr Israel Siegel, a biologist from sunny Florida where his invention should go down a treat, has come up with a container that works on the same principle as refrigerators and air conditioners. The difference is that those appliances use a chemical coolant called Freon, which is toxic and expensive. Dr Siegel uses a coolant that is non-toxic and cheap: it's called water.

When water is placed in a vacuum it has the same cooling properties as Freon, and is ideal for use with food and drink. Dr Siegel has simply surrounded a standard drinks can with a vacuum chamber, filled the vacuum with water and added a desiccant (a substance which absorbs water vapour from the vacuum chamber) on the outside. The cooling properties last for as long as the can remains unopened. Self-cooling cans will be about 25 per cent larger than standard ones and can be made from the same aluminium alloy.

FROZEN BEVERAGE MUG

●●● ✓ CONVENIENCE ✓ ENVIRONMENT-FRIENDLY

A lover of hot summer days, beautiful beaches and cold beer, Saul Freedman is the creator of the frozen beverage mug – an all-ice container that keeps the drink cold until the mug melts. Freedman's brainstorming idea – putting the liquid into ice rather than ice into liquid – was born out of frustration. He was tired of drinking warm beer and cola at the beach, and he was tired of seeing people's paper cups and cans littering the sand.

The frozen mug melts from the outside in (on a hot day it's good for almost forty-five minutes), and except for the (biodegradable) wooden stick that serves as its handle, it disappears without trace. It's produced by a patented process that first takes the impurities out of the water (making the ice freeze quickly and melt slowly) and then chills the moulds in super-cold storage.

Freedman is looking for investors, and then hopes to manufacture the mugs near his potential customers on the beaches of New Jersey. Franchising is a possibility, too.

BIODEGRADABLE PLASTIC BAGS

ENVIRONMENT-FRIENDLY

The plastic bags, bottles and other objects that are being ploughed into the ground at our refuse tips today will last four hundred years. Nor can we sensibly burn plastic rubbish, because that produces acid rain and damages the ozone layer. But help is on the way.

In 1989 the Italian government launched an anti-plastics campaign – all non-biodegradable packaging materials are to be phased out – so the incentive to speed up research was there. Sweden, too, is acting along similar lines, and EC funding is available to assist this kind of initiative. Now both Italian and US companies have come up with the world's first truly biodegradable plastics.

Most of the previously available so-called biodegradable plastics used starch, because it's biodegradable, to bind the plastic molecules into a sheet. But the good news stops there, because the remaining ingredients cannot be broken down.

The new plastics have managed to overcome this obstacle. Warner-Lambert, a New Jersey company, expects to start selling its product Novon, which consists entirely of starch and water, in a few years' time. Ferruzzi of Italy have developed a similar product with a high starch content; this one also contains an oil-based polymer that is water-soluble and therefore susceptible to normal bacterial breakdown processes.

The breakthrough has taken some time because up until now starch could not survive the heat and pressure of processing plastics without decomposing. Now researchers have harnessed starch's ability to melt under certain conditions before it decomposes, and have discovered how to injection-mould it like all thermoplastics materials.

An alternative technology being developed both in the USA and in Britain involves materials produced by living organisms such as bacteria. As a means of storing food, some bacteria produce the particular types of polymer from which plastics are made. But bacterially produced polymers are naturally biodegradable.

One of these polymers has rubber-like properties, and researchers think it might prove of use in health care. It would make very elastic, biodegradable surgical stitches; and because of that flexibility it might also be used by plastic surgeons as a kind of artificial skin to cover burns.

A third project, at the University of Chicago, is harnessing potato peelings and cheese scraps to make carrier bags that decay in sunlight or soil. Liquid discharges from potato and cheese processing factories provide a plentiful, cheap source of the necessary raw material, lactic acid.

The problem of what to do with our existing plastic detritus remains. But in the future, at least, we should not be adding to it.

FRESH FRUIT WRAP
✓ CONVENIENCE

Imagine perfect, unbruised, firm, fresh apples – as well as other fruits and vegetables – available any time of year in any part of the world. That's the vision of a group of American food science researchers at Cornell University.

The vision is well on the way to becoming reality, for the researchers have developed a prototype retail-sized food package that extends the storage life of apples for more than six months. A specially designed heavy-duty plastic tray is wrapped and sealed with a plastic film which controls the amount of oxygen and carbon dioxide passed to and from the apples. Other plastic wraps totally seal the apples and don't allow any passage of the gases.

Currently, fruit such as apples is stored in rooms with a mechanically controlled atmosphere. However, once these rooms are opened the apples rapidly begin to lose their eating quality. And you can't put any other kind of fruit in with them, because storage requirements differ.

With the new, high-tech packaging, on the other hand, different food products can be stored together because 'each package simulates a miniature controlled atmosphere storage system all its own', in the words of inventor Sayed S. H. Rizvi of Cornell. The plastic tray enables the contents to be stacked without getting bruised.

It is estimated that buying a dozen apples packaged like this will cost only slightly more than the current price of a bag of twelve. And the benefits will reach far beyond the shelves of your local supermarket. NASA has expressed interest in the new packaging, since it would enable astronauts to enjoy fresh fruit and vegetables during space flights.

FREEZE-DRIED COMPRESSED FOOD

●●○ ✓ CONVENIENCE ✓ SPACE-SAVING

It's compact, it's nutritious and it's tasty too. It's freeze-dried compressed food, a unique food preparation originally devised by the US military for submarines and space missions, where space (of the other sort) is at a premium.

What we're talking about is dehydrated food that can be reconstituted by adding water, like a sponge that swells when wet. The resultant bulk is what's important. With a compressed ratio of 16 to 1, a pound of freeze-dried compressed carrots, for instance, would give 16 lb of edible vegetables. 'You can take a chunk of beef stew the size of my thumb', says Dr Joseph Durocher of the department of hotel administration at the University of New Hampshire, 'and it will yield one and a half cups of stew.' However, only certain foods take well to the process. 'Veggies are great,' says Durocher, 'but you can't freeze-dry anything with a large mass, like a prime rib.'

The food is said to taste quite good, and as far as nutritional value is concerned it's better than tinned, but not as good as fresh. One advantage is that it has a very long shelf life.

Dr Durocher foresees his product being used some day by people living on space platforms and moon colonies. In the down-to-earth present, it is being used on a limited basis by campers.

NOISE-DRIED ORANGE JUICE

●●○ ☑ CONVENIENCE ☑ WACKY

Orange juice in powder form would certainly lighten your load from the supermarket. But dried by sound waves? Researchers from Purdue University in Indiana have done it, using a technique called acoustic drying that is based, bizarrely enough, on the engines that powered the German V1 rockets in World War II.

The new method is said to be cheaper than conventional drying methods, and to result in a better-quality product. Keeping the food or drink moving while it's being dried enables cooler air than normal to be used in the drying process, thus destroying fewer vitamins. This process will produce a product which will be as welcome to the supermarket manager as it will be to the consumer, for it takes up less space and has a longer shelf life.

Foods that are difficult to dry conventionally – corn syrup, soy sauce, apple juice and orange juice – have been dealt with successfully by the Purdue team. Unfortunately the engines used are extremely noisy. However, a new process is on the way that should be quieter for those in the vicinity, but just as effective.

POTATO ICE CREAM

●●● ✓ HEALTHIER

No, we're not talking about frozen chips – a company in Idaho, USA is actually making ice cream out of potatoes! And before you pass on with a shudder and a comment about 'nasty cheap substitute ingredients', consider this. Most of the milk solids are replaced with potato flakes that have gone through a special heating and cooling process that turns starch into fructose, nature's sweetener. The fructose is better for you than sugar (sucrose), and the potato base has half the calorie count of a comparable ice cream.

The product was the result of a tongue-in-cheek remark made to Alan Reed, president of Reed's Dairy, by his father. Alan wanted some more money pumped into his then all-dairy ice cream business. His father, who was on the National Potato Board, quipped that if Alan put some potatoes in the product, the board might consider it. Many a good idea has been born from odd circumstances; Alan Reed's potato ice-cream will soon be selling right across the United States. Perhaps somebody will take it up in Europe.

The ice cream, which has the taste and consistency of the more fattening kind, comes in traditional chocolate, vanilla and strawberry, as well as raspberry and orange cream.

HIGH-FIBRE CAKES

●●○   ✓ HEALTHIER

Is it possible that your favourite high-calorie, low nutritional value piece of sweet junk might one day actually be good for you? Some years ago, we all became aware from press and TV that certain forms of cancer might possibly be prevented by a higher fibre intake in our diet. At about the same time a team of American biochemists came up with a way of softening the non-digestible part of cell walls in cereals to make a cellulose fibre.

Experiments started to replace some of the flour content in baked goods with this new softened fibre product. The psychological strategy was well thought out. Nobody was going to double their fibre intake to avoid cancer some time in the future. The trick was to put the extra fibre into food that people liked eating – and to do it undetected.

'We made hundreds of cakes, brownies, doughnuts, pancakes and breads,' says Mike Gould, leader of the project. A tasting panel was asked to compare a standard cake with one in which 40 per cent of the flour had been substituted with cellulose fibre. They compared taste, texture, mouth feel and seven other criteria – and couldn't tell the difference! 'There was,' says Gould, 'as much fibre in one slice of that cake as in half a head of lettuce.' That was a quarter of the minimum daily amount recommended by cancer researchers.

Cellulose fibre can go in gravy, sauces, ice cream and any other products that require a bulking agent or a thickener, and it has no calories at all. It just passes right through the body.

NO-CALORIE SUGAR
●●○ ✓ HEALTHIER

This could be the breakthrough for which every chocoholic and junk food junkie has been waiting – sugars with hardly any calories.

They exist all right, and always have. They're called left-handed sugars as opposed to right-handed sugars – the ones we know and love. The two sugars have an identical molecular structure, except (as their names suggest) that they face in opposite directions. They are claimed to taste the same. The key difference is that the body's digestive enzymes only know how to break down the right-handed kind. The others get passed through the digestive tract completely unnoticed.

Left-handed sugars are found in minute quantities in rowanberries, seaweed, flaxweed gum, red algae and snails' eggs. Not the kinds of things we normally eat. And not in sufficient quantities to be farmed.

So we're left with the possibility of synthesizing these sugars in a lab, which is just what Dr Gilbert Levin, founder of Biospherics Inc. of Maryland, has done. He calls his product Lev-O-Cal, from Greek lev, meaning to rotate to the left; 'O' is for zero and 'Cal' for calories.

Unfortunately, until recently research and development took all the available money and nothing was left over for marketing. Then assistance turned up in the form of two Italian companies – Montedison and Eridania, the latter Europe's largest sugar firm. If current government testing goes well and there are no undesirable side-effects, Lev-O-Cal should be on the supermarket shelves by the middle of the decade, priced to compete with regular sugar.

And there's a plus point for those of us concerned with what goes into what we eat. Lev-O-Cal provides texture in baked goods, whereas other sweeteners need additives to serve as bulking agents.

NON-FATTENING FAT

●●○ ✓ HEALTHIER

Over the last few years we've been bombarded with exhortations to reduce our intake of animal fats and replace them with unsaturated vegetable oils and spreads, so as to cut down our risk of heart disease. Here's a more radical alternative – non-fattening fat: in other words, a product that looks, tastes and smells like real fat, but contains only a small fraction of the calories and has none of the dangerous effects on your arteries.

Procter & Gamble food scientists discovered a substance now named Olestra when they were looking for a way to improve the digestibility of fats and oils for people who needed to gain weight. They found that the molecules of this substance were too large to be absorbed by the body and therefore pass right through.

Testing is still being carried out, but the company hopes that, if approved, Olestra will replace 75 per cent of the fat in commercial deep-frying oils and salted snacks, and 35 per cent in home-cooking fats and oils. By mid-decade maybe that large portion of french fries needn't make you feel guilty any more.

CHOLESTEROL-FREE EGGS

●●○   ✓ HEALTHIER

Eggs that are 95 per cent cholesterol-free could be on your supermarket shelves very soon according to technologists at a Massachusetts company, Phasex. The key to this health breakthrough is a process called supercritical fluid extraction. In short, what scientists have found is that pure carbon dioxide dissolves the cholesterol from eggs. The technique is already being used by a German company to remove caffeine from coffee; it tastes excellent.

Cholesterol-free eggs will probably appear as an ingredient in baked products before we see the eggs themselves. These will be available in due course in powder or liquid form but are still being tested to ensure that the taste is retained and that they will be cost-effective to market.

CARBONATED MILK

●●○ ✓ WACKY

Milk drinking is now unfashionable in many European countries and the USA, while the consumption of soft drinks continues to shoot up. That's a worry for the dairy industry, so in a clever move that combines the best of both worlds the American United Dairy Industry Association is planning to launch carbonated milk drinks.

The new product will be found in supermarkets both with the dairy products and alongside nutritionally valueless: Coke, Pepsi, 7-Up and so on. If it tastes good enough, the marketing people reason, no one need know that it's actually healthy.

The drink will have a skim milk base (all the calcium, most of the vitamins and minerals, but none of the fat) to which fizz and flavour are added. Flavours under development include strawberry, chocolate, peach, orange, banana, pina colada, cola, rum and root beer.

Strangely enough, according to the food chemists, carbonated milk doesn't taste anything like the uncarbonated stuff. If it's unflavoured, it tastes like soda water; if it's fruit-flavoured, it tastes like a fruit drink.

SELF-STIRRING SAUCEPAN

●●● ✓ CONVENIENCE

Here's a neat little invention that should give cooks greater confidence and make their results more consistent. The French Tefal company has come up with an electric self-stirring saucepan that will de-lump gravy, gradually warm sauces, stir puddings and even scramble eggs.

Le Saucier Cook 'n Stir is non-stick and dishwasher-safe, and though it weighs a mere 3¾ lb it can hold 36 oz of liquid. A small motor in the base of the unit controls the stirring device, gently mixing ingredients as they are heated. The new pan comes with a recipe book that includes instructions on how to adjust the temperature according to what you're cooking.

MINI PORTABLE OVEN

●●● ✓ CONVENIENCE

The Porta-Oven could be described as a portable heating pad for food. It's actually a lightweight (less than 1 lb) heater-upper that can be used in cars, lorries, caravans, boats, at home – in fact anywhere that you can plug it in. There's an AC version for home and office use, and a DC one for boats and vehicles.

This handy object, slightly larger than a TV dinner, heats to 280° Fahrenheit but stays cool on the outside, and can handle anything from a pizza to a baby's bottle.

The envelope-like device is covered in a bright blue cloth and sealed with Velcro on the outside, and 'there's a state-of-the-art heating element inside' according to Kansas inventors Watkins Inc. The inner lining complies with government safety regulations, so food can be slipped in without any wrapping.

DEODORANT UNDERWEAR

●●○ ✓ CONVENIENCE

Attention all rugby players and whoever does their laundry! Here's an innovation that you'll welcome with open arms.

In Atlanta, Georgia, where textiles are king, a company called Interface Inc. has cooked up an anti-microbial agent that reduces the build-up of bacteria and fungus – the main causes of body odour. And then they found a way to add it to fabrics. The chemical used, which they call Intersept, is now being added experimentally to underwear, nappies and insoles for shoes.

The company say that the product 'has about the same amount of oral toxicity as table salt', but that doesn't mean that stringent testing won't be required before it becomes available to the public. So don't chuck out your roll-on yet – but just remember that the locker rooms of the year 2000 will never have smelt as good before!

ULTRASONIC 3-D CLOTHES

CONVENIENCE

Clothes manufacturers have gazed into their industry's crystal ball and seen this: garments moulded – not stitched – into 3-D shapes and produced in a mere forty-five seconds. That's where the industry should be before the mid-nineties. By the end of the century customers will choose colours, fabrics and details by computer, have body scans and get the finished outfit all in a matter of minutes.

Spearheading this new technology is Brett Stern, president of New York-based Symagery Inc. 'I realized,' he said, 'that there had been no improvement in the manufacturing process since the invention of the sewing machine.' But now, using the same technology that brought us 'the plastic cottage cheese container and moulded brassiere cups', he has invented a machine which 'takes a bolt of fabric and creates a garment in forty-five seconds without any human intervention'.

The cloth, made of at least 50 per cent synthetic fibre, is moulded into a 3-D shape, cut and ultrasonically sealed and finished along the seams. Exact size is of course guaranteed (there'll be no fiddly or time-consuming alterations), and 'by being made in a 3-D way, the clothing inherently fits the body better'. The garment will lie flat when finished, 'but will have a memory of the 3-D shape in which it was created'.

HOT/COOL FABRIC

CONVENIENCE

Imagine a lightweight jacket that keeps you as warm as a fur coat, and living room curtains that keep the air cool in a hot summer. Now imagine both of them in the same fabric. An impossibility? Not at all – an American process has now created a 'phase-change' fabric that has the ability to store, absorb and release heat.

A chemical called polyethylene glycol – it's used for temperature control in spacecraft – is applied to the fabric. When the temperature goes up, the molecules of the chemical absorb the heat and lock it in the cloth to keep you cool. When the temperature goes down, the fabric will warm you up by releasing the heat it's been storing in those molecules. The process is similar to the forming and melting of ice; at a certain temperature, the molecules take on different properties.

The treatment is being successfully applied in various ways to cotton, wood and synthetics. One limitation is that at present the fabric runs hot or cold for only half an hour at a time; scientists are busy trying to extend its staying power. But the treated material has proved to be more durable and wear-resistant than its untreated equivalent. And the chemical compound doesn't come out in the wash.

XEROGRAPHIC BEDLINEN

☑ CONVENIENCE ☑ ENVIRONMENT-FRIENDLY
☑ MONEY-SAVING

Imagine searching for bedlinen that not only matches your Laura Ashley chintz-covered chair but also highlights the sea green background of your antique rug. While shops are now much better at producing mix-and-match ranges and toning colours, you could still end up cross-eyed and footsore after searching through endless catalogues and tramping round every shop in town.

But the shopper of the future could find those elusive sheets and duvet covers in minutes. She – or he – would simply scan a computer display of patterns and press a few buttons; then an electrostatic printing system would begin printing the sheets. In a few more minutes the customer would be on the way home, bedlinen in hand.

The technology involved is called xerographic textile printing. It's a new application of the Xerox machines we're so familiar with in offices, and it's being developed in the Deep South, home of the American textile industry. Xerography works quickly and uses less energy than the traditional water-based dye process. Another advantage is that there's no dye-filled, polluted water left over.

There are still teething problems to be overcome. The polymers and pigments used with paper adhere quite well to cloth, but are less durable when exposed to sunlight, washed or worn. Colour registration is also not what it might be, creating a blurred image where it should be sharp – this is because cloth is less stiff than paper. But when all these are ironed out – as they surely will be – the new process could be a shot in the arm for the American textile industry. 'It could cut costs', says John Toon of the research team, 'and give US manufacturers a competitive edge over the Asian textile explosion.'

A NOTE ON SCIENTIFIC TERMS

If all you want to do is enjoy the sort of benefits of modern technology we've described in this book, then fine. But if you want to know a little more about some of the principles that lie behind the exciting innovations described, read on.

A *silicon chip* is essentially a sandwich made of three layers of silicon, a substance which conducts electricity moderately well – in other words it is a *semi-conductor*. The middle one of these three layers has had other substances added to it to remove its conductivity except when it receives a special electric current from the side. When this happens, electricity is able to pass right through all three layers. The device therefore acts as an electrically operated switch.

Solar power consists of light, heat and invisible radiation, all emanating from the sun. This power can be converted into electricity when it falls on a metal such as selenium; putting a conducting wire into contact with selenium enables the resulting current to be used. In solar heating the energy from the sun is converted from one form to another, for example from light to heat, or from invisible radiation to heat. Therefore bright sunlight is not necessary for the effect to work, which is why solar power can be harnessed even in colder northern countries.

A *hologram* means a complete picture, in the sense that it is a three-dimensional image produced on a flat surface of unexposed film. The image is created by bombarding the surface with a laser light beam (see below) which has been split into two. The first part is reflected from the subject of the hologram before falling on the film. The second beam is reflected directly on to the film. These beams consist of waves of light, but because of the way they are split they are offset when they hit the film – the front of one wave, for instance, hits the films when the middle of the other one is doing so. As a result, the two split beams interfere and create a pattern on the film which produces the 3-D effect picture in light and shade of the original subject.

Smart cards are plastic cards with built-in computer memory chips. These chips can store information which they release when the card is placed in a special reading machine. Among the thousands of possible applications the information extracted could, for example, include a unique password to open a security door for the card carrier.

A *laser* is an electrical device for generating intense light which can, if required, be focused into a very fine beam. It can be used for a variety of purposes which include producing displays of parallel beams of light extending over long distances, such as are used at rock concerts; or, because a fine beam can concentrate heat on to a small area, surgical cutting with a precision far greater than that of a scalpel.

A crystal normally has a rigid structure, unlike a liquid. *Liquid crystals,* however, are a halfway house. When an electrical current is passed through a liquid crystal, part of the structure rotates and reflects less light. These crystals can be made to take the shape of numbers and letters, for example in a liquid crystal display (LCD) in a calculator.

Fibre optics consists of the transmission of light along narrow 'wires' which are usually made of glass fibre. When these 'wires' bend, the light passing inside bends with them instead of continuing in a straight line. Among the uses of fibre optics is the medical examination of usually inaccessible parts of the body and, because many 'wires' can be clustered in a small diameter, for sending telephone messages in a space-saving form.

INDEX

3-D sound *83*

Anti-cavity pill *129*
Aquatic exercise machine *117*
Automatic tyre check and fill *167*

'Bark stopper' dog collar *45*
Binocular glasses *95*
Biodegradable plastic bags *184*
Butler-in-a-Box *46*

Camera stabilization lens *100*
Car video navigation system *158*
Car satellite navigation *159*
Carbonated milk *194*
CD ROM *64*

Cholesterol-free eggs *193*
Cholesterol-reducing powder *115*
Collision avoidance system *157*
Compact disc recorder *81*
Computer alarmcard *55*
Computer-designed dental crowns *131*
Computerized interior design *17*
Cool cooker *25*

Deodorant underwear *196*
Desktop video *89*
Diagnostic loo *152*
Dial 'M' for movies *94*
Digital hearing aid *125*

Digital implant for the totally deaf 126
Digital speakers 80
Digital audiotapes and decks 79
Dome homes 49
Drive by wire 162

Electrical shock hazard protector 40
Electrified birth control 147
Electronic still photography 98
Electronic bandage 140
Electronic supermarket 179
Electronic deodorant 141
Eye braces 120

Fever check 134
Fire emergency lifeline 39
Flat TV 86
Flexible bone clamp 135
Flying car 172
Four-wheel steering 169
Freeze-dried compressed food 187
Freezing humans 153

Fresh fruit wrap 186
Frozen beverage mug 183

Gyro exercise machine 116

Hand-scanning lock 42
Heads Up Display 163
High-fibre cakes 190
High-definition TV 85
Holograph bifocal contact lenses 118
Holographic phone 68
Hot/cool fabric 198

Impotence pills 149
In-flight entertainment system 175
Intelligent loo 27
Intercommunicating domestic appliances 20

Kiss moisturizer 142

Laminated wood to rival steel 19
Large-capacity smart cards 65
Larger-than-life TV 84
Laser treatment for sight

INDEX

defects *121*
Levitation vehicle *173*
Light-sensitive car windows and mirrors *165*
Line judge tennis balls *106*
Living skin equivalent *143*

Mammogram without X-rays *132*
Memory metals *39*
Memory card camers *99*
Microwave painkiller *138*
Microwave clothes dryer *23*
Mini portable oven *195*
Motion simulator *92*

Nail-less horseshoe *110*
Never-flat spare tyre *168*
Night vision display screen *164*
Night golf *102*
No-calorie sugar *191*
Noise canceller *31*
Noise-dried orange juice *188*

Noise meter *33*
Non-fattening fat *192*
Non-choking dog collar *44*
Non-invasive home glucose test for diabetics *133*

Oversized golf *101*

Painless dentist's drill *130*
Panic alarm wristwatch *76*
Perfect line golf ball *103*
Phone shields *151*
Plastic car engines *160*
Plastic nails *17*
Pocket organizer *54*
Pocket computer *53*
Portable voice-activated translator *73*
Potato ice cream *189*
Prayer wristwatch *75*
Pregaphone *136*
Privacy windows *37*
Private listening at home *82*
Punch meter *112*

Radon extractor *150*

Rear-view TV monitor system *166*
Remote control car starter *171*
Robot lawn mower *29*
Robot dog *43*

Safer tamper-proof packaging *182*
Seat bicycle pump *168*
Selection telephone *69*
Self-weeding lawn *31*
Self-stirring saucepan *195*
Self-cooling can *183*
Simulated golf *104*
Smart House *21*
Smart pill bottle *139*
Smart contact lens *124*
Smart TV *87*
Smart suspension *170*
Smoke Check badge *48*
Snore-reducing pillow *146*
Solar air conditioner *29*
Solar-powered briefcase *62*
Solar-powered cooker *24*
Solar roofing material *35*

Solar lighting system *34*
Solar windows *36*
Sonic painkiller *137*
Sports shock meter *111*
Spy satellite photos *97*
Stress gum *133*
Sunburn protection meter *142*
Super movies *93*
Super pore cleaner *144*
Super fish-eye lens *123*
Supermarket self-checkout *180*
Supermarket smart card *181*
Sweet-spot tennis racquet *105*
Swimming propulsion device *107*

Talking VCR remote control *88*
Talking glove for the deaf *127*
Telephone voice changer *70*
The ultimate computer notepad *58*
Time-release birth control *148*

INDEX

Time-release eye
   drops *119*
Touch-free taps *26*
Turbine car engines *159*
TV eyeglasses *122*
Two-cycle car
   engines *161*

Ultrasonic 3-D
   clothes *197*
Uphill skiing *109*

Video telephone *66*
Video-on-the-go *89*
Vilest taste *151*
Virtual world *90*
Voice-activated
   typewriter *71*
VoiceKey *41*

Walking Desk *63*
Walking TV *28*
Watch pager *74*
Water walkers *108*
Water battery *61*
Weather cube *60*
Window shatter
   protector *38*
World's smallest weather
   station *59*
Wrinkle-reducing
   pillow *145*

Write-top computer *56*

Xerographic bedlinen *199*